Praise for

THE FOURTH AGE

"*The Fourth Age* not only discusses what the rise of AI will mean for us, it also forces readers to challenge their preconceptions. And it manages to do all this in a way that is both entertaining and engaging."

—*The New York Times*

"Timely, highly informative, and certainly optimistic."

—*Booklist*

"In *The Fourth Age,* Byron Reese offers the reader something much more valuable than what to think about Artificial Intelligence and robotics— he focuses on HOW to think about these technologies, and the ways in which they will change the world forever. If you read just one book about the AI revolution, make it this one."

—John Mackey, cofounder and CEO, Whole Foods Market

"Reese frames the deepest questions of our time in clear language that invites readers to make their own choices. Using 100,000 years of human history as his guide, he explores the issues around artificial general intelligence, robots, consciousness, automation, the end of work, abundance, and immortality. As he does so, Reese reveals himself to be an optimist and urges us to use technology to build a better world."

—Bob Metcalfe, UT Austin Professor of Innovation,
Ethernet inventor, 3Com founder

THE FOURTH AGE

Smart Robots,
Conscious Computers,
and the
Future of Humanity

BYRON REESE

ATRIA PAPERBACK

New York · London · Toronto · Sydney · New Delhi

ATRIA
PAPERBACK

An Imprint of Simon & Schuster, Inc.
1230 Avenue of the Americas
New York, NY 10020

First Atria Paperback edition March 2020

ATRIA PAPERBACK and colophon are trademarks
of Simon & Schuster, Inc.

For information about special discounts for bulk purchases, please contact
Simon & Schuster Special Sales at 1-866-506-1949 or
business@simonandschuster.com.

The Simon & Schuster Speakers Bureau can bring authors to your live event. For
more information, or to book an event, contact the Simon & Schuster Speakers
Bureau at 1-866-248-3049 or visit our website at www.simonspeakers.com.

Interior design by Dana Sloan

Manufactured in the United States of America

1   3   5   7   9   10   8   6   4   2

Library of Congress Cataloging-in-Publication Data is available.

ISBN 978-1-5011-5856-8
ISBN 978-1-5011-5857-5 (pbk)
ISBN 978-1-5011-5858-2 (ebook)

To Sarah, Michael, John, and Peter,
who
every day
give me new reasons
to believe in
a better tomorrow

CONTENTS

PREFACE

(Please read. Not the usual blah-blah stuff.)

Robots. Jobs. Automation. Artificial intelligence. Conscious computers. Superintelligence. Abundance. A jobless future. "Useless" humans. The end of scarcity. Creative computers. Robot overlords. Unlimited wealth. The end of work. A permanent underclass.

Some of these phrases and concepts probably show up in your news feed every day. Sometimes the narratives are positive, full of hope for the future. Other times they are fearful and dystopian. And this dichotomy is puzzling. The experts on these various topics, all intelligent and informed people, make predictions about the future that are not just a *little* different, but that are dramatically different and diametrically opposed to each other. So, why do Elon Musk, Stephen Hawking, and Bill Gates *fear* artificial intelligence (AI) and express concern that it may be a threat to humanity's survival in the near future? And yet, why do an equally illustrious group, including Mark Zuckerberg, Andrew Ng, and Pedro Domingos, find this viewpoint so far-fetched as to be hardly even worth a rebuttal? Zuckerberg goes so far as to call people who peddle doomsday scenarios "pretty irresponsible," while Andrew Ng, one of the greatest minds in AI alive today, says that such concerns are like worrying about "overpopulation on Mars." After Elon Musk was quoted as saying "AI is a fundamental risk to the existence of human civilization,"

Pedro Domingos, a leading AI researcher and author, tweeted, "One word: Sigh." Each group's members are as confident in their position as they are scornful of the other side.

With respect to robots and automation, the situation is the same. The experts couldn't be further apart. Some say that *all* jobs will be lost to automation, or at the very least that we are about to enter a permanent Great Depression in which one part of the workforce will be unable to compete with robotic labor while the other part will live lavish lives of plenty with their high-tech futuristic jobs. Others roll their eyes at these concerns and point to automation's long track record of *raising* workers' productivity and wages, and speculate that a bigger problem will be a shortage of human laborers. While fistfights are uncommon between these groups, there is condescending invective aplenty.

Finally, when considering the question of whether computers will become conscious and therefore alive, the experts disagree yet again. Some believe that it's an obvious fact that computers can be conscious, and thus any other position is just silly superstition. Others emphatically disagree, saying that computers and living creatures are two very different things and that idea of a "living machine" is a contradiction in terms.

To those who follow all this debate, the net result is confusion and frustration. Many throw their hands up and surrender to the cacophony of competing viewpoints and conclude that if the people at the forefront of these technologies cannot agree on what will happen, then what hope do the rest of us have? They begin to view the future with fear and trepidation, concluding that these overwhelming questions must be inherently unanswerable.

Is there a path out of this? I think so. It begins when we realize that these experts disagree not because they *know* different things, but because they *believe* different things.

For instance, those who predict we will make conscious computers haven't come to that conclusion because they *know* something about consciousness that others don't, but because they *believe* something very basic: that *humans are fundamentally machines*. If humans are machines, it stands to reason that we can eventually build a mechanical human. On the other hand, those who think that machines will never achieve consciousness often hold that view because they aren't persuaded that humans are purely mechanical beings.

So that is what this book is about: deconstructing the core beliefs that undergird the various views on robots, jobs, AI, and consciousness. My goal is to be your guide through these thorny issues, dissecting all the assumptions that form the opinions that these experts so passionately and confidently avow.

This book is not at all about my own thoughts concerning these issues. While I make no deliberate effort to hide my beliefs, they are of little importance to how you, the reader, work your way through this book. My goal is for you to finish this book with a thorough understanding of where *your* beliefs lead you on these questions. Then when you hear some Silicon Valley titan or distinguished professor or Nobel laureate make a confident claim about robots or jobs or AI, you will instantly understand the beliefs that underlie their statements.

Where does a journey like this begin? By necessity, far in the past, as far back as the invention of language. The questions we will grapple with in this book aren't about transistors and neurons and algorithms and such. They are about the nature of reality, humanity, and mind. The confusion happens when we begin with "What jobs will robots take from humans?" instead of "What are humans?" Until we answer that second question, we can't meaningfully address the first.

So I invite you to join me on a brisk walk through 100,000 years of human history, discussing big questions along the way, and exploring the future to come. This book is a journey. Thank you for taking it with me.

Byron Reese
Austin, Texas

THE FOURTH AGE

INTRODUCTION

The most distinctive characteristic of the last century or so might seem to be the enormous amount of change that has occurred. Dozens, if not hundreds, of advances are said to have revolutionized our lives. The list includes automobiles, air travel, television, the personal computer, the Internet, and cell phones. Change is everywhere. We have harnessed the atom, flown into space, invented antibiotics, eliminated smallpox, and sequenced the genome.

But within the context of the overall arc of human history, little has changed in the past five thousand years. Just like the people who lived five millennia ago, we too have moms, dads, kids, schools, governments, religions, war, peace. We still celebrate births and mourn death. Forever with us, universal to all cultures of humanity, are sports, weddings, dancing, jewelry, tattoos, fashion, gossip, social hierarchy, fear, love, joy, happiness, and ecstasy. Looked at through this lens, humanity really hasn't changed much in all that time. We still go to work in the morning; only the way we get there has changed. In ancient Assyria, toddlers pulled around small wooden horsey toys on wheels with a string. In classical Greece, boys played tug-of-war. Ancient Egypt was renowned for its cosmetics, and millennia ago, Persians celebrated birthdays in much the same way as we do, with parties, presents, and special desserts.

No, the remarkable thing about our time is not the change we have

1

seen; rather, it is the change we *haven't* seen. The really amazing thing is how similar we are to our forebears. In ancient Rome, gladiators were paid celebrity spokesmen who recited product plugs just before the competition: "That's why I use Antinius's swords. You won't find a better sword at any price." And just like in our times, there were people willing to perform dramatically destructive acts just for the fame that doing so brought about, as was said to have happened on July 21, 356 BC, when an arsonist named Herostratus burned down the Temple of Artemis at Ephesus, one of the Seven Wonders of the Ancient World, simply for the everlasting fame it would bring him. In response, a law was passed that made saying his name a crime, but clearly Herostratus got his wish.

If you went to visit a friend in antiquity, you might have seen mounted on the door a brass lion's head with a ring in its mouth to be used to announce your arrival. If you attended a wedding five thousand years ago, you likely would have joined the wedding party as they wished good fortune on the new couple by throwing rice. Today, when we read that archaeologists have dug up ancient lead slingshot bullets each engraved with the word "catch," we still get the joke.

These people of antiquity were just like us. To really appreciate the unchanging nature of humans, one need look no farther than a book called *Characters* written by a Greek named Theophrastus 2,300 years ago. He satirized humanity itself, and sorted us all by type, such as the Flatterer, the Boor, the Chatty Man, and so forth. If you know someone who takes photos of his meals and posts them online, you might see him in the person Theophrastus calls the Garrulous Man, who "begins with a eulogy of his wife, relates the dream he had the night before, tells dish by dish what he had for supper," and concludes that "we are by no means the men we were" in times past. Theophrastus then goes on to describe the Stupid Man, who "when he goes to the play, is left at the end fast asleep in an empty house. . . . After a hearty supper he has to get up in the night, returns only half awake, misses the right door, and is bitten by his neighbor's dog."

No, against the backdrop of history, our time has seen very little change. In fact, I maintain that things have only *really* changed three times in human history. Each time was due to technology. Not just a single technology, but groups of interrelated technologies that changed us in fundamental and permanent, even biological, ways. That's it. Just three big changes so far.

This book is about the fourth one.

Part One

THE LONG, HARD ROAD
TO TODAY

THE STORY OF PROMETHEUS

The story of Prometheus is an ancient one, at least three thousand years old, and probably much older. It goes like this: Prometheus and his brother Epimetheus, both Titans (children of the gods, who came before the classical Olympian gods), were tasked with making all the creatures of the earth. Using clay, the brothers got to work. Epimetheus worked quickly making the animals, just slapping them together, and giving each of them one of the attributes Zeus had provided them to distribute. Some of the animals were thereby made cunning, some could camouflage themselves, some had fierce fangs, and some could fly. Prometheus, on the other hand, was a careful worker, and spent a great deal of time making just one creature—man—crafting him in the image of the gods, walking erect. By the time Prometheus was finished, he found out that his brother had given away all the gifts to the animals. One can almost picture Prometheus looking at the empty box that the gifts had been in, saying to his brother, "Dude. Really?" So Prometheus decided to do the one thing he was forbidden to do: give man fire. For this transgression, he paid a terrible price: he was sentenced by Zeus to be bound to a rock and to have an eagle pluck out his liver, which regrew every night, only to be torn out again the next day. This went on for eons until at last he was freed by Hercules.

1

The First Age: Language and Fire

While no one knows when isolated individual humans first harnessed the power of fire, we have pretty good evidence that around 100,000 years ago we gained widespread mastery of it. And it is easy to see in the much more recent Greek myth of Prometheus an eons-old memory of how dramatically fire changed us. Fire was the original multifunction technology. It provided light, and, because animals feared it, it also provided safety. Its portability meant that humans could migrate to colder climates and bring warmth with them. But far and away, its greatest benefit was that it let us cook food.

Why was this particular use of fire so important? Cooking allowed us to vastly increase our caloric intake. Not only does cooking meat make it easier to chew, but, more important, it unwinds the proteins within it, allowing for better digestion. And, on top of all that, with fire, scores of inedible plants suddenly became food sources, since fire could break down all the indigestible cellulose and starch found in them. Fire enabled us, in effect, to "outsource" part of our digestive process. It is quite difficult to get the number of calories humans require today with just raw food, since so many pass through the body undigested.

How did we use all the new calories we were able to consume? We used this new energy to grow our brains to unprecedented complexity. In a short period, we grew to have three times the number of neurons as gorillas or chimpanzees. Such a brain, however, is like an Italian super-car: it can go from zero to sixty in the blink of an eye, but it sure burns the gas doing so. Humans, in fact, use an incredibly lavish 20 percent of all of the calories we consume just to support our advanced brains. Few other creatures use even half that much energy to power their intellects. From a survival point of view, this was a pretty bold bet. To borrow a phrase from poker, humans went "all in" on the brain, and it paid off, for our more powerful brains led to our creation of another new technology: language. Language was the great leap that the historian Will Durant says "made us human."

So fire began the great romance we still have today with technol-ogy. What is technology? Throughout this book, when I use that term, I mean the application of knowledge to an item, process, or technique. And what is technology for? Primarily for enhancing human ability. It allows us to do things we couldn't do before, or allows us to do things we could do even better.

Certainly, we had used simple technology before fire. Over two mil-lion years before, in fact. But fire was different, special. It still seems a sort of magic. Even today, campers sit around the fire at night, staring into it, transfixed by its otherworldly dance.

The vastly more powerful technology of language allowed us to ex-change information. With it, you can encapsulate something you have learned, like "Tigers don't like to have their tails pulled," in a way that can efficiently travel from person to person, far beyond the one-handed man who had the original experience. Additionally, language enabled us to cooperate with each other, which is one of our singular abilities as a species. Without language, a dozen people were no match for a woolly mammoth. But with language, those people could work together in a way that made them nearly invincible.

Language came about because of our bigger brains, and in a virtuous cycle, language in turn grew our brains even bigger, as there are kinds of thoughts we cannot think without words. Words are symbols, after all, for ideas, and we can combine and alter those ideas in ways that are inconceivable without the technology of speech.

Another gift of language is stories. Stories are central to humanity, for they gave form to human imagination, which is the first requisite for progress. Oral chants, the progenitors of today's ballads, poems, and hip-hop songs, were probably early creations of speaking humans. There is a reason that things that rhyme are more memorable than things that don't. It is the same reason that you can remember song lyrics better than a page of prose. Our brains are wired that way, and it is that fact that allowed *The Iliad* and *The Odyssey* to be preserved in oral form before the invention of writing. This also explains why the opening theme songs for TV shows like *Gilligan's Island*, *The Beverly Hillbillies*, and *The Brady Bunch* are forever etched into my mind, in spite of my not having seen those shows for decades. It is notable that those songs are stories themselves, even including the words "story" and "tale" in their lyrics. It is speculated that our oldest stories, like the epic of Gilgamesh, probably existed in oral form for millennia, until the invention of writing allowed them to be jotted down.

We don't know much about our earliest language other than what we can infer from our languages today. The original tongue of humans is long gone, as are the many forms that came after it. We group present-day languages into families that are derived from theorized proto-languages. One such language is Proto-Indo-European, from which 445 languages today descend. These include Hindi, English, Russian, German, and Punjabi.

Philologists study protolanguages by looking at similarities between words across languages. In 2013, researchers from the University of Reading in England employed this sort of analysis to find the oldest words we use. Their research found twenty-three "ultraconserved" words that have likely sounded largely the same for fifteen thousand years, meaning they reach back to a protolanguage that existed even before Proto-Indo-European. These most ancient of words include "man," "mother," "two," "three," "five," "hear," "ashes," and "worm." The oldest word of all may be "mama" or something like it, given that in a large variety of languages, the word for mother begins with the *m*-sound, often the first sound a baby can make.

And then, intriguingly, we have languages that seem to have no linguistic antecedents, languages that seem to have come out of nowhere. Basque, spoken by the people who inhabit the mountains between Spain and France, is one such example. It is thought by many to be a language older than the Proto-Indo-European tongue, and there is a legend among the Basques that theirs is the language that was spoken by Adam and Eve in the Garden of Eden.

The versatility and complexity of language are amazing. English recently passed a million unique words, although most of us get by using only about twenty-five thousand. A new English word is coined approximately every hour, although that pace is slowing. Back in the day, someone like Shakespeare would make up three words before breakfast. The leading theory as to the culprit behind this recent slowdown in new words is automatic spell-checkers, which simply won't put up with that kind of horseplay. And unless you want to send an email with lots of red scribbles under your so-called words, you had better use stuff already on the approved list.

During this time, the First Age (the roughly 100,000 years in which we lived as hunter-gatherers with both language and fire), what was life like? The total population of humans was around 200,000, so while we were not an endangered species, the survival of humanity was far from certain. Although there was undoubtedly a wide variation of practices, a large number of these people lived in collectivist, largely nonhierarchical societies. As recently as 1700, there were still over fifty million hunter-gatherers spread across the globe, so we have a good deal of firsthand observations about "modern" hunter-gatherers. Even today, the best estimates suggest that there exist more than a hundred uncontacted hunter-gatherer tribes, whose members may number a total of over 10,000 people.

If modern examples of hunter-gatherers are any indication of life before agriculture, we can infer that sustenance was not something one could take for granted, and that any individual was just a few days' illness away from death. As such, a general collectivism likely arose from each individual having a compelling, self-interested motive for helping others: even the strongest members of a society would need help themselves someday. For this reason, groups that shared were likely to be more resilient than their more selfish brethren. Besides, what was the point of accumulating wealth? There *was* no wealth beyond the day's haul of grubworms and no way to store wealth even if it was there. Humans lived day to day, eking out a meager existence, just one bad winter or rogue mammoth away from an untimely demise.

Modern-day followers of Rousseau have a tendency to look back on this time through the rose-colored glasses of romanticism, harking back to a simpler time, when humans lived in harmony with nature, uncorrupted by the trappings of the modern world. Most of us, if dropped back into that time to live out our days, would likely not conclude these were the good old days. To begin with, times were violent. The Harvard psychologist Steven Pinker estimates that, based on studies of ancient human remains, nearly one in six ancient hunter-gatherers met with a violent end at the hands of another human. Compare this with just the one in thirty who died such a death in the "bloody" twentieth century, with its two world wars. Thus we can confidently say that life as an ancient hunter-gatherer was short, painful, and harsh. But this was humanity's proving ground, and with language, we embarked on the path that brought us to today.

2

The Second Age: Agriculture and Cities

After about 100,000 years of humans chatting away while they hunted and gathered their way through the day, something dramatic happened that profoundly altered humans and our society once again: we invented agriculture. The Second Age began just 10,000 short years ago, when the human population of the planet was about four million people, a little more than the current population of Los Angeles. In those 90,000 intervening years, we managed to double our population a mere four or five times. That is incredibly meager growth, indicative of the precariousness of our existence.

Agriculture, like language, is also a technology, and like language, agriculture brought about a slew of other advances. The first of these was the city, which came about as agriculture required that humans settle down in one place. This practice was almost entirely new. Early cities, such as Çatalhüyük, Jericho, and Abu Hureyra, were often located near rivers for access to water and fertile farmland, and had markets, homes, and temples. It was during the Second Age that we began using opium, gambling with dice, and wearing makeup and gold jewelry.

Cities promoted commerce and the exchange of ideas, but they also made us completely and irrevocably stationary. Housing was permanent. We altered the land through dikes and terraces. We built fences. And we buried and marked the location of our dead for later homage. These practices, and dozens more, were nails in the coffin of our wandering nature. There would be no turning back.

The second technological advance to come along with agriculture was the division of labor. While this may not seem like such a complex idea, the impact of the division of labor marks *the* major milestone in the history of humanity. With the division of labor, instead of each individual doing all of the things necessary for survival, individuals specialize in narrower tasks, and by doing so gain efficiencies, which allow for tremendous economic growth. Along with trading and technological advances, the division of labor is one of only three "free lunches" in conventional economic theory—that is, one of the ways overall wealth can be increased without anyone's having to work any harder.

Agriculture didn't give us the division of labor directly. It gave us cities, and cities gave us the division of labor. How? The division of labor works the best when large numbers of people live in close proximity. Farmers who may have lived far from their neighbors couldn't really specialize, and thus were by necessity jacks-of-all-trades but masters of none. Imagine how unproductive you would be if, instead of your current job, which you are presumably pretty good at, you had to do everything for yourself, from sewing your own clothes to making your own soap. Archaeological evidence from the oldest cities suggests that there was a range of different jobs from the beginning of the Second Age. Humans reaped the incredible economic advantages of specialization the moment that they started living in close proximity with large numbers of other humans.

The division of labor makes cooperation between humans go from optional to required. The famous essay by the economist Leonard Read, "I, Pencil," describes how although no one person knows how to build a simple pencil, the pencil still gets made, because thousands of people in hundreds of fields, who will never meet, each do a small part of each thing that it takes to make a pencil. The division of labor gives us virtually everything we have today. Without it, we would perish.

Weapons for organized warfare are another technology that came along because of the city. They were invented out of necessity, because the city concentrated wealth and needed to be defended. The earliest cities were often walled, which is achieved only through great effort and expense, implying that the risks of invasion were real, or at least perceived to be.

As a result of agriculture and cities, humanity had individual private ownership of land for the first time. Humans, being territorial, have probably always defended a loosely defined area they regarded as their own, but we have archaeological evidence from the beginning of the Second Age that borders were often well defined. The philosopher Jean-Jacques Rousseau thought of this practice as the beginning of our modern world, and stated that "the first man who, having fenced in a piece of land, said 'This is mine,' and found people naïve enough to believe him, that man was the true founder of civil society."

Agriculture and privately owned land ended the economic equality of the First Age. The natural inequality of ability, birth, and luck led to unequal accumulations of wealth. Although coinage in the modern sense didn't exist at this time, the idea of wealth certainly did. One could own land, cattle, and silos for storing grain. That wealth could be accumulated indefinitely, with no upper limit on how rich a person could be. Since land could be farmed and cattle could reproduce, early wealth was income producing. As such, holdings of wealth tended to grow. Given that wealth could be passed down from generation to generation, it could accumulate and compound over multiple lifetimes.

Sadly, it was in the Second Age that the practice of human enslavement began. Slavery made little financial sense in a hunter-gatherer world, where wealth was nonexistent or at the most ephemeral, lasting only a day or two. But with cities, land ownership, and stores of wealth, our innate acquisitiveness was kindled, and was further stoked by memories of times of privation. The hunger for wealth seems to be limitless, at least in some, as is evidenced by those who feverishly work to earn their second billion, even with the full knowledge that they won't ever, in a hundred lifetimes, spend their first.

Slavery presented no real ethical challenge for a world that didn't have a notion of human rights or individual liberty. Only later, as civilization progressed, did the immorality of the institution become glaringly obvious.

Over time, some people amassed more land and capital than others. As society became wealthier, complexity arose. Trade became more sophisticated. Technology advanced and cities grew. All of this together raised the upper limit of the amount of wealth a single person could accumulate.

An unintended consequence of the agricultural revolution was that while more food could be produced, food could also be withheld from people. In a hunter-gatherer world, that wasn't really possible, but with cities and agriculture, withholding food was a way for those in power to silence opposition, while distributing food was a way to ensure obedience. This is still done in parts of the world today.

It is against this backdrop that people separated into the rulers and the ruled. Aristocracy and royalty emerged during the Second Age. The ruling classes frequently adopted practices that those they ruled over were not permitted, such as wearing certain types or colors of clothing, eating certain foods, or, in the case of the Aztecs, smelling certain types of flowers.

This is also where the tension between two values, freedom and equality, was first highlighted. As the historian Will Durant points out, you get to pick only one, because you can't have them both. People truly free will become unequal. People with equality forced on them are not free. This tug-of-war still plays out today.

Earlier, I referred to imagination as the first requisite for progress. Agriculture gave us the second. Since planting and harvesting crops required planning in a way that hunting and gathering did not, we can think of the invention of agriculture as the invention of the idea of the future, which is the second requisite for progress.

3

The Third Age: Writing and Wheels

Fire let us cook food, giving us our brains, which in turn produced language, allowing us to work together, form abstract thoughts, and create stories. Ten thousand years ago, agriculture let us settle down, build cities, and accumulate wealth. Cities were fertile grounds for the division of labor to produce economic growth and innovation.

The Third Age began just five thousand years ago when writing was likely first invented by the Sumerians, a people who lived in the southern part of present-day Iraq. It seems also to have been developed independently at around the same time in Egypt and China; some scholars give the "earliest writing" award to the Chinese. It would later be developed independently in what is modern-day Mexico. Writing changed humanity because for the first time what a person knew could live after him or her, perfectly preserved. Knowledge could be flawlessly copied and transported around the world. Ideas could live outside a human mind!

None of these benefits was the cause of the invention of writing. Writing in its earliest forms was about keeping track of assets and transactions. From there, it spread out to cover legal records, legal codes, and religious texts. Creative writing such as plays and poetry would come later.

At first, very few of the planet's ten million people could read, as would be expected. Widespread literacy was slow in coming, due to the high costs associated with writing. Not only did it take a huge time investment to learn to read and write, but "paper" was either papyrus, baked clay, marble, or other unforgiving and expensive media.

But the power of writing caused it to quickly branch out into all parts of life, and in doing so, transformed the world. What of our modern world would exist without this technology? Writing is the great dividing point in human history. The First and Second Ages are, by definition, prehistoric. History begins five thousand years ago with the Third Age.

Not everyone thought writing was a good idea. In his writings, Plato describes a great king rebuking the god who invented writing, telling him:

> For this invention will produce forgetfulness in the minds of those who learn to use it, because they will not practice their memory. . . . You have invented an elixir not of remembering, but of reminding; and you offer your pupils the appearance of wisdom, not true wisdom.

Plato was right. Writing does hurt our memories. Just as we use fire to outsource some of our digestion, writing outsources our memory. Before writing, if you wanted to know something, you had better remember it, because there was no way to write anything down. Ancient history brings us tantalizing hints of a time when we had better memories, while I struggle to remember my ATM PIN. But our memories didn't degrade immediately with the invention of writing, because books were still uncommon. Now that most knowledge is a Google search away, our memories may further decay.

Like the other pivotal technologies we have explored, writing also had concurrent new technologies it helped bring about or promote. The first of these was the wheel, which came along at the same time, about five thousand years ago. The wheel and writing go together like PB and J, for as a pair they increased commerce, aided the flow of information, and promoted travel. Writing meant that rulers could create legal codes, but it was the wheel that allowed those codes to be distributed and enforced across a wide area.

Early legal codes were so sparse that in more than one culture, the ruler was required to memorize every law before taking power. It is from this period that the maxim "Ignorance of the law is no excuse" came about, for since there were so few laws, you had no excuse not to know them. Although we still say this, practically speaking, the exact opposite is true: Ignorance seems like a pretty good excuse in a country with millions of pages of laws. Early legal codes, such as the four-thousand-year-old Code of Ur-Nammu, specify specific punishments for murder, robbery, kidnapping, rape, perjury, assault, and a variety of crimes relating to land ownership, such as flooding a neighbor's field, failing to cultivate a rented field, and secretly cultivating someone else's field. The Code of Hammurabi, only a couple of centuries younger, covers all that ground in its 282 laws, but it adds in the enforcement of contracts, product liability, and inheritance.

Money appeared during the Third Age as well. Stamped coins as we have today wouldn't be developed until well into the period, but money in a dozen other forms, from gold and silver to shells and salt, appeared all over the world early in the Third Age. Metals were considered the ideal media of exchange because they were widely valued, divisible, durable, and portable. Metallurgy itself began at the dawn of the Third Age, and humans soon learned that tin and copper combined to produce bronze, which is superior to both.

With writing, the wheel, and money all coming on the scene concurrently, the basic ingredients needed to make the nation-state and empires were in place. This is when we saw the first large civilizations blossom all over the world, independently and virtually simultaneously. China, the Indus Valley, Mesopotamia, Egypt, and Central America all became home to large, cohesive, and prosperous nations. No one has any idea how it came to be that these civilizations emerged nearly concurrently in places with no contact with each other. The same is true for writing. Why didn't some parts of the world get writing, wheels, agriculture fifty thousand years ago? Or twenty thousand? No one knows.

So by this point in our telling of human history, we have language, imagination, division of labor, cities, and a sense of the future. We have writing, legal codes, the wheel, contracts, and money. All of this together allowed us to advance our technology relatively quickly over the next few thousand years.

Our world up to recent times has been a Third Age world. While incredible innovation has occurred along the way, such as the development of the steam engine, the harnessing of electrical power, and the invention of movable type, these were not fundamental changes in the nature of being human in the way language, agriculture, and writing were. The signature innovations within the Third Age have been evolutionary more than revolutionary. This is not to diminish them. Printing changed the world profoundly, but it was simply a cheaper and more efficient way to do something that we already could do. Detailed schematics of a biplane would have made sense to Leonardo da Vinci. For us to rightly say we have launched into a new age, something has to have come along that has changed us and how we live in a profound and permanent way. Something that has altered our trajectory as a species.

And that story, which launches us into the Fourth Age, has its origins in the final few centuries of the Third Age.

4

The Fourth Age: Robots and AI

Although we are accustomed to technology advancing rapidly, for more than 99.9 percent of human history, it didn't. One of the oldest tools, the Acheulean hand ax, remained unchanged across a million years of use. Imagine that! Unchanged for a million years. Today's technology advances at a brisker pace, but it has been doing so for only a few centuries. Some historians say that Leonardo da Vinci was the last man to know everything. And while this isn't meant in a literal sense, it is an acknowledgment that Leonardo lived at a time when science was so nascent that a single person could have a working knowledge of all that was known.

But by the time Leonardo died in 1519, things had already begun to change. By the middle of that century, Nicolaus Copernicus reordered the cosmos in *On the Revolutions of the Heavenly Spheres*. Shortly thereafter, a French philosopher named Jean Bodin was one of a group of people who saw science as the way forward. Bodin didn't believe in some golden age of the past; rather, he believed that the power of print would launch the world forward and that the sciences "contain in themselves treasure that no future ages will ever be able to exhaust."

By 1600, things were really getting under way. In 1609, Johannes Kepler wrote a letter to Galileo Galilei talking about a future with space-ships: "Ships and sails proper for the heavenly air should be fashioned. Then there will also be people who do not shrink from the dreary vastness of space." In 1620, Sir Francis Bacon published a book called *Novum Organum* (*The New Method*), which is regarded as the beginning of what we now call the scientific method. Bacon emphasized the firsthand study of nature along with careful observations and the recording of data. From that data, and only from that data, should one draw conclusions.

While this isn't exactly how we think of the scientific method today, Bacon was important because he proposed a way to systematize the acquisition of knowledge through observation. That's a big idea, a world-changing idea. For up until this time, progress came in fits and starts, as the wheel was both literally and metaphorically reinvented again and again. With the scientific method, the data and conclusions that one person collects can later be used by others to advance knowledge further. This enabled a compounding growth in our scientific knowledge, which is what got us to today.

Today's scientific method is a set of agreed-upon techniques for acquiring knowledge, and then distributing that knowledge in such a way that others can corroborate and build on top of it. It applies only to objects or phenomena of which measurement can be made. Objective measurement is essential because it allows others to reproduce a researcher's findings, or, as is often the case, to be unable to reproduce them. The scientific method required affordable printing to work properly, which is probably why it wasn't developed earlier in human history, and why science advanced ever faster as the cost of printing fell.

The ancients had many extraordinary technological breakthroughs, but since they lacked the technology and a process for publishing and disseminating information about them, they were quickly forgotten. One example is the Antikythera mechanism, a two-thousand-year-old

Greek mechanical device (a computer really) that was used to forecast astronomical positions and calculate when eclipses would occur. We know of this device only because we found exactly one example in a shipwreck. In our modern world, such a revolutionary device would be written about and photographed ad nauseam. Colleges around the world would try to outdo each other making improvements to the device. Entrepreneurs would be raising money to build Antikythera mechanisms cheaper, smaller, and faster.

That's how technology advances: by making incremental improvements on work done by others, a process Isaac Newton described as seeing further by standing on the shoulders of giants. In 1687, Newton, on whose shoulders we still stand, published *Philosophiae Naturalis Principia Mathematica*, which describes the laws of motion and gravitation. In just a few formulas, Newton showed that even the planets themselves obey straightforward, mechanistic laws.

It would be an oversimplification to give all the credit for our rapid technological advance to the scientific method. That was simply the last piece of a complex puzzle. As I've already pointed out, we had to have, among other things, imagination, a sense of time, and writing. In addition, we needed much more; to that list we might well add a low-cost way to distribute knowledge, widespread literacy, the rule of law, nonconfiscatory taxation, individual liberty, and a culture that promoted risk-taking.

The invention of the printing press, and its widespread use, increased literacy and the free flow of information. This was the main catalyst that launched our modern world way back in the seventeenth century. And, perhaps, modernity got an unexpected boost from something else that happened in Europe at the same time: the replacement of beer by the newly introduced coffee as the beverage one sipped on all day. Now that bottled water is the beverage du jour, we might inadvertently slip back into a new dark age. But at least we will be well hydrated. Or, conversely, maybe Starbucks will save the world.

The scientific method supercharged technological development so much that it revealed an innate but mysterious property of all sorts of technology, a consistent and repeated doubling of its capabilities over fixed periods.

Our discovery of this profound and mysterious property of technology began modestly just half a century ago when Gordon Moore, one of the founders of Intel, noticed something interesting: the number of transistors in an integrated circuit was doubling about every two years. He noticed that this phenomenon had been going on for a while, and he speculated that the trend could continue for another decade. This observation became known as Moore's law.

Doubling the number of transistors in an integrated circuit doubles the power of the computer. If that were the entire story, it would be of minor interest. But along came Ray Kurzweil, who made an amazing observation: computers have been doubling in power from way before transistors were even invented.

Kurzweil found that if you graph the processing power of computers since 1890, when simple electromechanical devices were used to help with the US census, computers doubled in processing power every other year, regardless of the underlying technology. Think about that: the underlying technology of the computer went from being mechanical to using relays, then to vacuum tubes, then to transistors, and then to integrated circuits, and all along the way, Moore's law never hiccupped. How could this be?

Well, the short answer is that no one knows. If you figure it out, tell me and we can split the Nobel money. How could the abstraction, the speed of the device, obey such a rigid law? Not only does no one really know, there aren't even many ideas. But it appears to be some kind of law of the universe, that it takes a certain amount of technology to get to a place, and then once you have it, you're able to use that technology to double that again.

Moore's law continues to this day, well past the ten years Moore himself guessed it would hold up. And although every few years you see headlines like "Is This the End of Moore's Law?" as is the case with almost all headlines phrased as a question, the answer is no. There are presently all manner of candidates that promise to keep the law going, from quantum computers to single-atom transistors to entirely new materials.

But—and here is the really interesting part—almost all types of technology, not just computers, seem to obey a Moore's law of their own. The power of a given technology may not double every *two* years, but it doubles in something every *n* years. Anyone who has bought laptops or digital cameras or computer monitors over time has experienced this firsthand. Hard drives can hold more, megapixels keep rising, and screen resolutions increase.

There are even those who maintain that multicellular life behaves this way, doubling in complexity every 376 million years. This intriguing thesis, offered by the geneticists Richard Gordon and Alexei Sharov, posits that multicellular life is about ten billion years old, predating earth itself, implying . . . well, implying all kinds of things, such as that human life must have originated somewhere else in the galaxy, and through one method or another made its way here.

The fact that technology doubles is a big deal, bigger than one might first suspect. Humans famously underestimate the significance of constant doubling because nothing in our daily lives behaves that way. You don't wake up with two kids, then four kids, then eight, then sixteen. Our bank balances don't go from $100 to $200 to $400 to $800, day after day.

To understand just how quickly something that repeatedly doubles gets really big, consider the story of the invention of chess. About a thousand years ago, a mathematician in what is today India is said to have brought his creation to the ruler, and showed him how the game was played. The ruler, quite impressed, asked the mathematician what he wanted for a reward. The mathematician responded that he was a humble man and his needs were few. He simply asked that a single grain of rice be placed on the first square of the chessboard. Then two on the second, four on the third, each square doubling along the way. All he wanted was the rice that would be on the sixty-fourth square.

So how much rice do you think this is? Given my setup to the story, you know it will be a big number. But just imagine what that much rice would look like. Would it fill a silo? A warehouse? It is actually more rice than has been cultivated in the entire history of humanity. By the way, when the ruler figured this out, he had the mathematician put to death, so there is another life lesson to be learned here.

Think also of a domino rally, in which you have a row of dominos lined up and you push one and it pushes the next one, and so on. Each domino can push over a domino 50 percent taller than itself. So if you set up thirty-two dominos, each 50 percent bigger than the first, that last domino could knock over the Empire State Building. And that is with a mere 50 percent growth rate, not doubling.

If you think we have seen some pretty amazing technological advances in our day, then fasten your seat belt. With computers, we are on the sixtieth or sixty-first square of our chessboard, metaphorically, where doubling is a pretty big deal. If you don't have the computing power to do something, just wait two years and you will have twice as much. Sure, it took us thousands of years to build the computer on your desk, but in just two more years, we will have built one twice as powerful. Two years after that, twice as powerful again. So while it took us almost five thousand years to get from the abacus to the iPad, twenty-five years from now, we will have something as far ahead of the iPad as it is ahead of the abacus. We can't even imagine or wrap our heads around what that thing will be.

The combination of the scientific method and Moore's mysterious law is what has given us the explosion of new technology that is part and parcel of our daily life. It gave us robots, nanotech, the gene editing technology CRISPR-Cas9, space travel, atomic power, and a hundred other wonders. In fact, technology advances at such a rate that we are, for the most part, numb to the wonder of it all. New technology comes with such rapidity that it has become almost mundane. We carry supercomputers in our pockets that let us communicate instantly with almost anyone on the planet. These devices are so ubiquitous that even children have them, and they are so inexpensive as to be free with a two-year cellular contract. We have powers that used to be attributed to the gods, such as seeing events as they happen from a great distance. We can change the temperature of the room in which we are sitting with the smallest movement of our fingers. We can fly through the air six miles above the Earth at the speed of sound, so safely that statistically one would have to fly every day for over 100,000 years to get in an accident. And yet somehow we can manage to feel inconvenienced when they run out of the turkey wrap and we have to eat the Cobb salad.

In antiquity if you wanted to know the answer to a hard question, you would have to make a pilgrimage to see an oracle, such as the Oracle at Delphi. After a long, arduous journey full of many hardships, you would finally present your query to the oracle, who would reply, in a drug-induced stupor, some cryptic answer that could be interpreted a dozen different ways. Contrast that now to Google: you type in a question, and in a quarter of a second, Google rank-orders fifty billion web pages for your perusal.

Of all the technologies that have come our way in the final years of the Third Age, hands down, one advance tops them all: the computer. Computers are not simply gadgets. They are philosophically meaningful devices. Why do I say this? It is because they do one very special thing, computation. I know that seems screamingly obvious, but computation is the heartbeat of the universe, it is the ticktock of the cosmic clock. Computation is so foundational that some believe that *everything* is computation: your brain, the universe, space, time, consciousness, and life itself. The polymath Stephen Wolfram thinks so, and he makes his case across 1,200 pages in his 2002 magnum opus *A New Kind of Science*. He demonstrates that very simple rules, just one or two lines long, can generate immense complexity. He further speculates that the code required to generate the entire universe may just be a few lines long. It is a provocative hypothesis, one with many adherents.

Regardless, much of the universe clearly is computational. Hurricanes and DNA are computational, as are snowflakes and sand dunes.

The wondrous truth is that things that are the result of computation in the physical world can be modeled inside a computational device the size of a postage stamp. Just think about it for a minute. Putting a man on the moon involved incredibly complex calculations about rockets and boost and gravity . . . all stuff of the real world. And yet we could mirror it with ones and zeros inside a tiny processor. This suggests a profound truth, that anything that can be modeled in a computer is occurring via computation in the real world as well. In other words, Apollo 11's launch was computational. It didn't merely involve computation, its launch *was* computation. Its entire mission was computation.

Where it gets interesting are the edge cases. Are we computation? Are our minds giant clockworks that follow the same basic rules as Apollo 11 did? Those are the questions that we need to answer in order to understand the limits, if there are any, of computers.

This is why I say that computers are philosophically significant devices. A hammer just pounds nails, a saw just cuts wood, but a computer can mirror a billion different things from the physical world. It is fair to say that we don't yet understand the metaphysical implications of the computer. We know that it has transformed the world in ways both subtle and dramatic; that's obvious. But more is going on than meets the eye. As the noted professor and philosopher Marshall McLuhan said decades ago, the computer is "the most extraordinary of man's technological clothing; it's an extension of our central nervous system. Beside it, the wheel is a mere hula-hoop." The computer is both new and ubiquitous at the same time, and one can only imagine what it will able to do in a century, or even a decade.

Where did this device come from? How is it that we decided to make one, or even suspect such a thing was possible? The history from the beginning of computers to today is quite short, and for our purposes here can be told by mentioning just four names, which when listed together sound like a high-tech law firm: Babbage, Turing, von Neumann, and Shannon. Let's look at just one idea from each, which, when taken together, will give you the essentials of modern computing.

The story begins in 1821 in London with Charles Babbage. At the time, the Industrial Revolution was under way, and science and math moved from the university and the laboratory into the factory. In the days before calculating machines, large books of tables were published that could be used as shortcuts by those doing complex calculations. These books contained logarithms, astronomical calculations, and loads of other data sets essential for industry and science. The problem was that every number in those books was hand-calculated, and thus they contained numerous errors. One wrong number could send a ship off course, corrupt bank records, or cause the production of faulty machinery. Babbage, frustrated by the errors, remarked, "I wish to God these calculations had been executed by steam."

This statement was profound, and an extremely modern one at that. Captured in it is a sense that things mechanical are more consistent and reliable than things organic. Steam-powered machines were carefully milled to exacting standards. Those machines performed tasks indefatigably, producing goods of consistently high quality. It was a stroke of genius on Babbage's part to realize that if steam could make cogs, it could also compute logs.

Thus he conceived of and tried to build a complete calculating machine. He understood the significance of the machine, stating that "as soon as an Analytical Engine exists, it will necessarily guide the future course of science." Unfortunately, he ran out of funds for his endeavor, a common fate for start-ups even then. However, in 2002 the Science Museum of London built the ten-thousand-pound machine Babbage proposed, and it worked flawlessly. Exit Babbage, who surmised that steam could power computing machines.

Enter Alan Turing. Turing's contribution at this point in our tale came in 1936, when he first described what we now call a Turing machine. Turing conceived of a hypothetical machine that could perform complex mathematical problems. The machine is made up of a narrow strip of graph paper, which, in theory, is infinitely long. On the graph paper there is always a single active cell, and above that cell hovers a head. The head can read and write to the paper and move around a bit, based on the instructions it receives or the programs it runs.

The point of the Turing machine was not "Here's how you build a computer" but rather, "This simple imaginary device can solve an enormous range of computational problems. Almost all of them." In fact, anything a computer can do today, you could theoretically do on a Turing machine. And Turing not only conceived of the machine but figured all this out. Consider that simple machine, that thought experiment with just a handful of parts: Everything Apollo 11 needed to do to make it to the moon and back could be programmed on a Turing machine. Everything your smartphone can do can be programmed on a Turing machine, and everything IBM Watson can do can be programmed on a Turing machine. Who could have guessed that such a humble little device could do all that? Well, Turing could, of course. But no one else seems to have had that singular idea. Exit Turing.

Enter John von Neumann, whom we call the father of modern computing. In 1945, he developed the von Neumann architecture for computers. While Turing machines are purely theoretical, designed to frame the question of what computers *can* do, the von Neumann architecture is about *how* to build actual computers. He suggested an internal processor and computer memory that holds both programs and data. In addition to the computer's memory, there might also be external storage to hold data and information not currently needed. Throw in input and output devices, and one has a von Neumann setup. If, when you were reading that, your brain mapped it to your computer's CPU, memory, hard drive, keyboard, and monitor, then move to the head of the class.

Finally, in 1949, Claude Shannon wrote a paper entitled "Programming a Computer for Playing Chess" in which he described a way to reduce chess to a series of calculations that could be performed on a computer. While this may not sound like it should earn Shannon one of the four spots on the Mount Rushmore of computer history, for the first time, in a practical and realistic way, computers were thought of as not just machines to perform mathematical calculations. Shannon allowed that they could manipulate information at an abstract level, as necessary to make chess moves. Think about that: Before 1949, computers were programmable calculators, the kind you need in physics class. After 1949, it was conceivable that a computer might someday suggest which stocks you should buy. While many others, Turing included, intellectually knew what computers were capable of, Shannon delivered the goods.

So that's that: Babbage realized machines could do math. Turing added that they could also run programs. Von Neumann figured out how to build the hardware, and Shannon showed how the software could do things that didn't at first look like math problems.

That is where we are today. The only things that have really changed are that computers have gotten a whole lot faster and whole lot cheaper. One way to see this phenomenon is to examine the cost of a computer that can do a gigaflop, that is, one billion floating point operations per second, and ask what's the cost per gigaflop. In 1961, no such computer existed, but if you took two years of the United States' GNP, spent it all on the fastest computers of the day, and bolted them together, you would get close to one gigaflop.

By 1984 the price had plummeted, and you could get a one-gigaflop computer, a Cray "supercomputer," for about the price of buying a really nice private jet—a bargain compared to two years of GNP. By 1997, you could get one gigaflop of computing power for the price of a good German sports car. By 2013, the price had fallen to just twenty-five cents per gigaflop, and that supercomputer was called a Sony PlayStation 4. Now the price is around a nickel, as today's PCs are ten-thousand-gigaflop machines and cost just a few hundred dollars.

The price per gigaflop of computer power will be a fraction of a cent soon enough, and will continue to speedily drop from there. Supercomputers today aren't measured in gigaflops anymore, nor even in teraflops (1,000 gigaflops), but in petaflops (1,000,000 gigaflops). The fastest computer as of January 2018 is located in China. As is the second fastest. They operate at nearly 100 petaflops (100,000,000 gigaflops). However, later in 2018, computers in excess of 100 petaflops will be built in the United States, Japan, and undoubtedly elsewhere. In addition, there are no fewer than five companies promising an exaflop (1,000 petaflops) machine by 2020. There is no hint that this race will slow down any time soon.

How did all this happen? How does the price fall like that? In 1960, you could buy a transistor for about a dollar, or just over $8 in today's money. So, if you needed 125,000 of them, you needed a million dollars in today's money. But the price fell dramatically as the quantity produced skyrocketed. By 2004, the number of transistors manufactured surpassed the number of grains of rice that were grown across the planet. Just six years later, in 2010, you could buy that same 125,000 transistors that cost a million dollars in 1960 for the same price as you would pay for a single grain of rice.

Technology is relentless: It gets better and cheaper, never stopping. And it is on this fact that many computer scientists base their claims on the future capabilities of computers, such as artificial general intelligence and machine consciousness, the topics we will discuss for the rest of the book.

Just how deeply have we embedded computers into the fabric of our lives? No one knows how many billions of computers are in operation around the world. It is believed that computers use roughly 10 percent of all of the electricity produced. They are so much a part of our lives that we literally may not be able to live without them, certainly not at our present standard of living. But our population may be large enough that without computers in the background managing everything from logistics to water treatment, their removal or incapacitation might cause a die-off of humans, especially in large cities. As Steve Wozniak said, "All of a sudden, we've lost a lot of control. We can't turn off our Internet; we can't turn off our smartphones; we can't turn off our computers. You used to ask a smart person a question. Now who do you ask? It starts with g-o, and it's not God."

In the 1960s and 1970s, we were building enough computers that it made sense to connect them to make one giant network. We call that the Internet. In 1989, Tim Berners-Lee created a protocol called HTTP to access a document on a server from a remote computer. We call this the World Wide Web. Today we are in the process of connecting not just computers to the Internet, but every device that is driven by data. Thirty billion or so devices are connected now, and that number is expected to rise to half a trillion by 2030.

That is the story that gets us from the waning centuries of the Third Age to the doorstep of the Fourth Age. Each new age saw technology outsourcing and enhancing functions of our physical or mental life. We used fire to help with digestion, writing to augment our memories, the wheel to spare our backs and legs. In our time, we have created a device, a mechanical brain, that is so versatile that it can be programmed to solve virtually a limitless number of problems that we ask it to. We are now developing artificial intelligence, a method to teach that device to operate on its own, and through the power of robotics, we have begun to give it mobility and ways to interact with the physical world. We will use computers and robots to outsource more and more of our thoughts and actions, presumably as much as we possibly can. This is a real change, and it marks the dawn of a new age, the Fourth Age. The questions that come out of this transition are profound, for they relate to what it means to be human. Can machines think? Can they be conscious? Can all human activity be mechanically reproduced? Are we simply machines? The ultimate purpose of this book is to explore these ideas—to figure out just how much human activity, both mental and physical, we can delegate to machines, and what the implications of that change will mean for the world.

Has the Fourth Age begun yet? Well, when just a few humans learned to farm, or a few isolated places developed writing, did that mark the beginning of a new age or the beginning of the end of an old one? It matters very little where we draw the line. Whether it already happened decades ago when Claude Shannon explained how a computer could be programmed to play chess, or whether it will happen in a few years when a computer can carry on a complex conversation with a human using natural language, this kind of hairsplitting is not particularly meaningful. Let's say that the transition began no earlier than 1950 and will be completed no later than 2050. Draw your line anywhere in there you choose, or label the entire period as a transition.

When it began doesn't really matter, for the important thing to understand is that once it really gets going, change happens rapidly. It took us 5,000 years to get from the wheel to the moon. (Interestingly, we would make it to the moon before it occurred to anyone to attach a wheel to luggage.) But we weren't halfway to the moon 2,500 years ago. Far from it. When a human first walked on the moon, we had broken the sound barrier only two decades earlier. We had achieved heavier-than-air flight only six decades earlier. So for 4,940 years after inventing the wheel, we were still firmly attached to the face of this planet. Sixty years later, we hadn't just flown, we'd gone to the moon and back. It is at that speed, with frequent dramatic transformative breakthroughs, that you should expect this fourth change to descend on us. And just as airplanes experience the most turbulence right before landing, our ride from this point forward is going to be a bit bumpy. We will probably see more change in the next fifty years than we have seen in the last five thousand. Vladimir Lenin said, "There are decades where nothing happens; and there are weeks where decades happen." As the Fourth Age gains steam, expect breakthroughs in AI and robotics to happen with ever-increasing rapidity.

For millions of years, we have used technology. For a hundred thousand years, we have had language. A few thousand years ago, we started asking profound questions about the universe and our place within it. A few centuries ago, we systematized science and created unimagined prosperity. A few decades ago, we began making mechanical brains, and just a few years ago, we learned dramatic new ways to make those brains more powerful. It feels as if this very moment is the great inflection point of history.

Now we have arrived at the doorstep of the Fourth Age, and we find ourselves in disagreement about just exactly what we have built and where we are heading. The technologies we are now creating will force us to reexamine our answers to some very old questions.

5

Three Big Questions

Now that the Fourth Age is upon us, we are ready to tackle the issues of robots, jobs, artificial intelligence, and conscious computers. To do so, I need to pose to you, the reader, three foundational questions. I will be referring back to them repeatedly throughout the rest of this book. They are old questions, ones that we have been asking ourselves for thousands of years. It would be inaccurate to say, "There are no wrong answers," for without a doubt, there are many wrong answers. There just isn't agreement over what they are.

Why are these questions important? They are the issues on which hinge much of what we are about to discuss. These include what artificial intelligence will be capable of, whether computers can become conscious, and if robots will take all our jobs. While one might expect us to jump into technical details, that's not the best way to approach those questions. If conscious computers—machines that think, and robots that duplicate the full range of human ability—are possible, then the steady advance of technology will eventually deliver them to us. In other words, if they are possible, then they are inevitable. But are they in fact possible? This question is far from settled, and it depends entirely on the answers to these three questions.

The questions are philosophical, not technical; you don't need expertise in AI or robotics to answer them. Historically, these questions haven't had much in the way of real-world implications. They were mainly academic mental exercises, the kind of thing that you might discuss with your college friends late at night. But how that has changed! At the dawn of the Fourth Age, they have suddenly become immensely practical questions. In a very real sense, the answers to these three questions will determine the future and the fate of our entire species. I will repeat myself: what were for thousands of years simply abstract and, and to many people, largely irrelevant philosophical ponderings are now the central questions of our time. If one knows their answers, the future becomes much clearer. If tackling the great philosophical issues seems daunting to you, then, well, the only consolation I can offer is that they are presented as multiple-choice questions.

What Is the Composition of the Universe?

The first question is about the composition of the universe, the very nature of reality. However you answer this question will go a long way to determining whether computers can be conscious and whether true artificial intelligence can be built. The best thinking about this question comes to us from the ancient Greeks. It's pretty unusual that we have a situation in which there exists about 2,500 years of debate centered on the exact same texts. Since we haven't bettered the ancient Greeks' reasoning by much in the intervening millennia, we tackle this first question by going back to them and how they thought about it long ago.

What is the composition of the universe? There are two schools of thought.

The first is that everything in the universe is composed of a single substance, atoms. This is known as monism, from the Greek *monos*, meaning "one." Monists believe that everything in the universe is governed by the same set of physical laws, and that those laws are largely known to us today. Nothing in the universe happens that cannot ultimately be reduced to physics. Physics sits on top of everything. Physics explains chemistry, chemistry explains biology, biology explains life, life explains consciousness. Monism is also called materialism or physicalism.

One ancient Greek representative of this position is Democritus, who believed that the only things that existed were matter and empty space, and he classified everything else as simply opinion. He said:

> *By convention sweet is sweet, by convention bitter is bitter, by convention hot is hot, by convention cold is cold, by convention color is color. But in reality there are atoms and the void. That is, the objects of sense are supposed to be real and it is customary to regard them as such, but in truth they are not. Only the atoms and the void are real.*

A modern proponent of monism is Francis Crick, who offered his "astonishing hypothesis" that "you, your joys and your sorrows, your memories and your ambitions, your sense of personal identity and free will, are in fact no more than the behavior of a vast assembly of nerve cells and their associated molecules."

A great majority of scientists identify with this viewpoint. In fact, to most scientists this position seems screamingly obvious, although they often have an understanding of the three reasons why this view is unsettling to many people:

First, it is hard to work free will into a world of simple cause and effect.

Second, it means we are nothing more than big walking bags of chemicals and electrical impulses. Despite what your mother told you, there is nothing special about you at all. You are the same basic "thing" as an iPhone, a turnip, or a hurricane.

Third, it is hard to coax a universal moral code out of that viewpoint. Killing a person doesn't seem to have any more moral consequence than smashing a boulder.

There are of course plausible answers to these points, which in turn invite plausible challenges to those answers. And that is where the debate begins.

The other school of thought is a position known as dualism. Dualists believe that the universe is made of two (or more) things. Yes, there are atoms, but there is also something else.

There is a temptation to spiritualize this position, while concurrently casting monism as the rational, modern view. And while it is true that those who believe in God or the soul or ghosts or "life forces" are certainly dualists, the dualism tent is much larger and includes many viewpoints that eschew the spiritual. Atheism and theism are beliefs about God; monism and dualism are beliefs about the nature of reality.

What might these other substances be? There are a couple of ways to think about it. One is that there is physical stuff and spiritual stuff. There are atoms and there are souls, for instance. This is the more religious conception of dualism. The second way to think of dualism doesn't have any religious implications. It is that the universe is composed of physical stuff and mental stuff. The mental stuff includes hopes and regrets, love and hate, and so forth. While these things may be triggered by physical processes in the brain, the *experience* of them is not physical. It is a subtle difference, but an important one.

We will use Plato as our ancient Greek advocate for dualism. He believed that while there are things we call circles in this world, they are never really *perfect* circles. But there exists a form, an ideal, that is a perfect circle, and that "circle-ness" is a real thing that isn't governed by physical laws. And yet with your mind, you can visit the world of perfect forms by, for instance, contemplating philosophy. In a sense, high school geometry is an exercise in Platonic dualism. Geometric proofs rely on perfect forms, but in reality, there are no true circles, lines, or planes. Geometric proofs only loosely resemble our reality.

A modern Platonist might say, "You know what an idea is. It is a noun, right? You *have* an idea. It is a thing." Now, here's the question: Does an idea obey the laws of physics? Some would say yes, that the idea is a pattern of neurons in your brain. In your brain, it displays entropy. But some would say no, that ideas are things untouched by physical laws. They originate at some place and time, they spread, they have an effect on their environments. But physics doesn't govern ideas.

The modern defense of dualism is best expressed by the "Mary's room" problem, formulated by the philosopher Frank Jackson. It goes like this: Mary is a hypothetical person who *knows everything* about color. Literally everything. Not just about the science of color, but right down to how photons hit the eye, and what the cones and rods do. Not just what they do, but at an atomic level, what is going on. I mean, she knows every single thing about color there is to know. But she has spent her entire life in a gray room reading about color on a gray screen. She has never actually seen any color. But remember, she *knows* everything about it.

Then one day she leaves the room and sees color for the first time. Now, here is the big question: Does she now know anything more about color? Did she learn anything after seeing color for the first time? If you believe she did, if you think that experiencing something is different than knowing something, then you are a dualist. If she learned *anything* new, then it means that there is something that happens in experience that is beyond the physical universe, beyond simply knowing a thing. Whatever Mary learned about color when she saw it for the first time is something that is outside the realm of physics. What is that thing? How would you express it in an equation? Or even words, for that matter? Holding this belief will have profound implications on your conceptions of what a computer can and cannot do.

Another modern proponent of dualism is René Descartes, of "I think, therefore I am" fame. (You know you are dealing with an ancient problem when a guy born more than four hundred years ago is considered "modern.") A quick refresher on his famous statement. Descartes began by doubting everything. Does two plus two equal four? Maybe. Does sewer rat taste like pumpkin pie? Who knows? He doubted the entire material world under the idea that it might be a trick being played on him by a demon of some kind. The conclusion of all this doubting is that . . . well, he knew nothing, other than that he doubted everything. So that was his foundational conclusion, better rendered as "I doubt, therefore I am." Descartes was the quintessential dualist, and he saw the mind as the source of consciousness and the brain as the place where simple facts were. So in his view, consciousness was mental and knowledge was physical. We will explore the difference between the brain and the mind later.

The traditional argument against dualism asks this question: If there is a world of physical things and a very different world of mental things, how do they interact? How can having a mental desire of wanting a sandwich compel your physical body to get up and make one? The fact that the craving can influence your body suggests that they are both physical.

All this may sound like sophomoric hairsplitting, but it will become quite germane when we ask whether computers can, for example, ever "feel" pain. They can sense temperature, but can they experience pain?

So, I now ask you to call the ball on our first question. Are you a monist or a dualist? (If you think you are something other than those two things, consider yourself a dualist for our purposes. The salient point is that you are not a monist.)

What Are We?

Next question: What exactly are we? Again, a multiple-choice question, with three possible answers: machines, animals, or humans.

The first possible answer is that we are machines. This is the simplest, most straightforward answer. We are a bunch of parts that work together to achieve an end. We have a power source and an exhaust system. We self-repair and can reprogram ourselves to do a variety of different tasks.

Those who hold this viewpoint are quick to caution against thinking of the term "machine" as pejorative. We may "just" be machines, but we are the most amazing and powerful machine on the planet. Maybe in the universe. Your basic essence may be the same as a clock radio, but your form is so much more wondrous as to make the comparison ridiculous except in a purely academic sense.

Those who hold this belief maintain that everything that happens in your body is mechanistic. That is true almost by definition. It is neither miracle nor magic that keeps your heart pumping. We are simply self-sustaining chemical reactions. Your brain, while not fully understood, gives up more of its secrets every day. In a lab, an imaging device can already read some of your thoughts. If someone built an atom-by-atom copy of you, it would show up at your office tomorrow with a packed lunch, ready to work. You could easily get away with sneaking out the back door and going fishing, because that copy of you would do exactly the same job you would. Come to think of it, it would probably take the day off and go fishing, knowing that you were going into the office.

This viewpoint calls to mind a thought experiment articulated best by the philosopher Derek Parfit. You have probably considered something like it before. In the future there is a teleportation device. You step in. It takes you apart, painlessly, cell by cell, scanning each cell. The data about that cell is beamed to Mars, where a similar device does the opposite: it builds someone cell by cell who is identical to you in every way. That person steps out and says, "Man, that was easy."

Would you step into such a device? The majority of people probably would not consider that "person" on Mars to be themselves. They would probably consider it a creepy doppelgänger of themselves. And yet it is incumbent on them to explain just exactly what attribute they have that a high-enough resolution 3-D scanner can't capture. But to people who believe they are machines, there is nothing philosophically troubling about such a device. Why would you ever want to wait in traffic when you can just step into the teleporter?

With regards to life, this view holds that it too is simply a mechanistic process. Consciousness as well. To those who hold this view, all this is painfully apparent, and they do not wince when they read Kurt Vonnegut's thoughts on this question:

> *I had come to the conclusion that there was nothing sacred about myself or any human being, that we were all machines. . . . I no more harbored sacredness than did a Pontiac, a mousetrap, or a South Bend Lathe.*

Your second choice is that we are animals. Often this view sees an inorganic, mechanical world that is a completely different thing from the biological, living world. Life makes us different from machines. Maybe our bodies are machines, but "we" are animals that inhabit those machines.

This position maintains that there is something to life that is more than electrochemical, for if it were solely that, we could build a living thing with a couple of batteries and a sufficiently advanced chemistry set. Life has some animating force, some mysterious quality that perhaps is not beyond science, but is beyond machines. New stars are born, crystals grow, and volcanoes die. But although these objects exhibit these characteristics of life, we don't think they are alive. Machines, in this view, are the same sort of thing, lifelike but lifeless.

This distinction between living animals and nonliving machines seems a natural and obvious one. While we anthropomorphize our machines and talk about "the car not *wanting* to start because the battery is *dead*," we use those words without contemplating therapy for the car or mourning the loss of the battery whose life was cruelly cut short.

Life is something we don't fully understand. We don't even have a consensus definition of what it is. If you do believe that we are animals, and that we are different from machines because we are alive, then our big question is going to be if computers can become alive. Can something that is purely mechanical get the spark of life? We will get to that in part four.

The final choice is that we are humans. Everyone agrees that we are called humans; I mean something more here. This position says that of course our bodies are machines, and yes, of course we are alive like animals. But there is something about us that separates us from the other machines and animals, and makes us a completely different thing. We aren't just the ultimate apex predator, the preeminent animal on the planet. We are something fundamentally different. What makes us different? Many would say that it is either that we have consciousness or that we have a soul. Others say it is that we make and use complex tools or that we have mastered complex language or can reason abstractly. Maybe humanity is something emergent, some by-product of the complexity of our brains. Aristotle suggested that what makes us human is that we laugh. The Dalai Lama expressed it as such: "Humans are not machines. We are something more. We have feeling and experience. Material comforts are not sufficient to satisfy us. We need something deeper—human affection."

We share a huge amount of DNA with every living thing on the planet, including plants. This notion is profound, and it is one best expressed by the author Matt Ridley in just four words: "All life is one." Beyond this idea of unity, we share as much as 99 percent of our genome with a single species: chimpanzees. So as machines and animals, we are strikingly similar to chimps, with only a rounding error of difference. But viewed through another lens, we are absolutely nothing like chimps. And whatever that lens is, *that* is what makes us human. However, just having some *difference* from animals doesn't make us not an animal. The distinguishing factor has to be something that changes our essential selves. For instance, humans are the only creatures that cook their food. But that distinction alone doesn't make something more than an animal. If we suddenly discover a magpie in Borneo that drops crabs into fires and retrieves them later to eat, we wouldn't grant "humanness" to magpies. But if that same magpie developed a written language and began writing limericks, well, we would have to consider it. So that is the question: Is there something about us that makes us no longer only an animal?

Interestingly, some of the Greeks divided the living world into those three similar categories as well. Plants, the reasoning went, have one soul, because they are clearly alive, and eat, grow, reproduce, and die. Animals have two souls. That same one the plants have, but another one as well: they are purposeful. Finally, there are humans, who have three souls. The plant soul, the animal soul, and a third one, a reasoning soul, because only we can reason.

So it is decision time: Are we machines, animals, or humans?

What Is Your "Self"?

So now we come to our third and final foundational question: What is your "self"? When you look in a mirror and see your own eyes, you recognize yourself in there. What is the thing that looks back at you? What is that voice in your head that talks to you? What is the "I" that you mean when you say, "Oh, I understand"?

We have a sense of feeling that our "self" lives in our heads not for any real biological reason, but because in our modern era we associate the "I" with the brain. But the sense that you can feel yourself thinking in your brain is probably an illusion. How do we know? Well, before modern times, people felt like they were thinking with different parts of the body. The Egyptians, for instance, who saved all of a person's body parts during mummification because he or she would need them in the afterlife, threw the brain out as useless, thinking it was just a goo that kept the blood cool. Aristotle had this same view of the function of the brain. In other cultures, people have thought the self lived in the heart, with cognition occurring there. That is why we learn something "by heart" and we love someone "with all our heart." Given the centrality of the heart to life, this makes a kind of sense.

In other places and times, anatomists who noticed the centrality of the liver to the body's various systems located the self there. We still have vestiges of this in our language as well, such as when we know something "in our gut" or have a "gut reaction" to something.

To further probe this question before we get down to possible answers, ask yourself what makes you, you. When you look at a photo of yourself as a baby, in what sense is that you? When you wake up in the morning, what makes you the same person you were the night before? Is it your memories? If that is the case, then a person who gets amnesia no longer exists. Perhaps the continuity of your physical form makes you, you. Whenever that possibility is posed, the thought experiment of the ship of Theseus is sailed in.

I will give you an abbreviated version: There is a famous ship in a museum. As pieces rot over the centuries, they are replaced. This happens so much that eventually no original piece of the ship remains. Is it still the same ship? The additional wrinkle to the problem is that all the rotted pieces have been saved in a storeroom and someone pieces them all back together to form a derelict ship of Theseus. What do we say then? There are two ships of Theseus? The point of the whole thing should be obvious: you are a breathing ship of Theseus. Given that your cells replace themselves, you literally aren't the same matter you were a decade ago. But is that you? Given that your brain cells don't regenerate, or regenerate relatively little, it might be tempting to say that you are your brain cells. However, while the cells don't regenerate, they constantly change in their relationship with each other. So it is hard to pin down exactly what "you" are.

Why does this question matter for our purposes here? Well, whatever your "self" is, it is inextricably tied to consciousness. It is hard to conceive of a consciousness without a self. So in later chapters we will be keenly interested in whether computers can have a "self."

Let's tackle our question, then: What is your "self"? There are three possible choices: a clever trick of your brain; an emergent mind; or your soul.

The first option is that it is a trick of your brain. This is what most brain scientists believe. "Trick" in this sense is not intended to mean a deception, but a clever solution, like, "I know a trick for getting gum out of hair."

So what is this trick? It has two parts. First, your brain gets inputs from all kinds of senses. You receive images through your eyes, you sense temperature with your skin, you hear things with your ears, and so forth. But you don't perceive reality that way. You don't have to integrate those different sensory inputs consciously, right? Your brain has figured out this cool trick whereby it can blend them into one single mental experience. It combines them all together. You see the rose and smell it at the same time. It is all one integrated experience, even though sight and smell are completely different parts of the brain. That is the first half of the trick.

Now the second half. At the same time that all this is going on, the different parts of your brain are chugging along, doing their respective things. Some parts are watching out for danger and others are doing math and still other parts are trying to remember the lyrics to a song, and so forth. And the brain has figured out that the best way to handle all this noise is to let just one part of the brain at a time "have the floor." So maybe you are sitting in a coffee shop alone, trying to remember the lyrics to "Louie, Louie" and in walks a grizzly bear who looks like he has had a very bad day. All of a sudden, the part of your brain that is looking for danger starts yelling, "*Bear! Bear!*" and that part forcefully takes the floor, pushing aside all thoughts of the Kingsmen as you scan the room for weapons or exits. That ability to switch focus is the second part of the trick. Instead of hearing a constant cacophony in your head akin to the mayhem that used to happen on the floor of the New York Stock Exchange before computers took over, you get a nice, orderly single voice, as different parts of the brain take turns "speaking." Have you ever had the experience of trying to remember something, then it's just popping into your head an hour later? That's because the part of your brain looking for it never stopped doing so; it just receded into the background while it worked and grabbed the floor only when it finally remembered.

So that is all that the "you" is. Your brain puts all the senses together into a single show and lets only one part speak at a time. Those two things together give the illusion of a "you."

The way I have described this process, some part of the brain would seem to be in charge, a cerebral master of ceremonies, picking who gets to speak and what you see. If that was the case, then that part of your brain would be the "you." But most brain scientists believe that such a part of the brain doesn't exist. The brain self-regulates. There is no executive "you" in charge. Think of it as a giant cocktail party. There is all this chatter going on. Then all of a sudden, a woman yells, "My dress is on fire!" and the whole place turns and faces her. A man grabs a vase, yanks the flowers out, and douses her with water. Meanwhile, another guest takes his jacket off and starts beating the fire out. Just as that ruckus is settling down, someone else yells, "Is that Elvis outside?" and everyone runs over to look out the window. In those cases, no one was in charge. Nobody was calling the shots, different processes were just commanding attention.

In this view of your "self," you would have experienced the voice in your head saying, "That woman's dress is on fire," followed by that voice saying, "Could that really be Elvis?" According to the "trick of the mind" choice, that's the entirety of what your "self" is.

The second option is that your "self" is an emergent mind.

Emergence is a fascinating phenomenon. At its simplest, emergence is when a group of things interact with each other, and through that interaction, the collective whole gains characteristics that none of the individual things has.

Humans are clearly emergent things. You are made of forty trillion cells. They all go about their daily business doing their jobs, getting married, having kids, and then dying. And all along, they have absolutely no idea you exist, or that they are even part of something else. But you, and all your abilities and attributes, are not simply the abilities and attributes of a single one of your cells multiplied by forty trillion. You are not simply the cumulative result of your biological processes, the sum of your individual parts. Not a single one of those forty trillion cells has a sense of humor, and yet somehow you do. Somehow there is an "I" that arises from all the activity of those forty trillion cells doing their respective things. We call that emergence. While we understand that it happens, and that in a sense it powers the universe, we don't really understand *how* it happens.

There is a kind of giant honeybee that exhibits an emergent property called shimmering. Now, your average giant honeybee isn't really that bright. In fact, a genius giant honeybee isn't bright either. But collectively they do something pretty smart. Can you picture that spinning spiral pattern that is always used in cartoons to hypnotize someone? Well, imagine a clump of honeybees the size of a dinner plate making that pattern by turning their dark abdomens up at just the right instant, similar to how the wave is done in stadiums. But imagine it spinning a few times a second. There is no way one honeybee sees another one doing it, then thinks, "Now it is my turn." They just spin and spin, and do so to scare away hornets, who seem to be totally freaked out by this display. But there is no honeybee who is in charge who signals the others to fire up the shimmer.

Or take ant colonies. Any given ant is even less intelligent than a bee. And yet the colony does incredible things, such as building nests, excavating tunnels, and responding to changes in the weather. There are all these different ant jobs that need to be done, and individual ants leave one job and take up another. Balances are struck between ants staying back to protect the colony and ants going out to get food. And if there is a picnic going on nearby, some ants will leave one job to go get more food. But here is the thing: no ant is in charge. The queen isn't, she just lays the eggs. No ant tells any other ant what to do. The colony *itself* can thus be thought of as having an emergent mind.

This choice is quite different from the prior one. The "trick of your mind" view suggests that brain function is basically understood and is reasonably straightforward, with all the appropriate caveats. The voice in your head is simply different parts of your brain grabbing the microphone. The emergent view is that things are going on in your brain that are much less understood. Your mind emerged from the basic parts of your brain and took on characteristics that are not as simply explained as the "trick of your mind" view holds. If you are fine with a nuts-and-bolts view of the brain as a straightforward, albeit utterly amazing, organ, then you might find the "trick of your mind" view to be most appealing. If you struggle with a notion that that view doesn't really account for creativity or you want to insist that there is a real you that is calling the shots, you may find the emergence camp to be more to your liking.

The final option is that your "self" is your soul. The majority of people probably believe this. Why do I say that? Religious belief, while not universal, is certainly the norm. Poll after poll after poll shows that an overwhelming (75 percent or more) portion of Americans believe in God, the devil, heaven, hell, miracles, and the soul. Around the world, we see the same story. While belief in God is not as high worldwide as it is in the United States, it doesn't dip much below a majority in any country. Worldwide, the best estimates are that 75 percent of people believe in God, 15 percent are agnostic, and 10 percent are atheists.

Meanwhile, the percentage of Americans who believe in Darwinian evolution unguided by the hand of God rests at 19 percent, a fact that causes no small amount of bewilderment and frustration to those 19 percent.

Since the sensation or belief that one has a soul is an entirely private experience, completely real to the person who experiences it but invisible to everyone else, it is a challenge for science to deal with. And while someday science may determine if the brain works the way I describe earlier in this chapter and come to understand emergence better, the soul, by definition, exists outside the physics of the material world, and thus is subject to neither proof nor disproof that would pass scientific muster. Of course, this is fine, because I am sure that almost all those reading this already have some opinion one way or another as to whether they have a soul, and they're not waiting for a scientific journal to weigh in.

So what is your essential "self": A trick of your brain, an emergent property, or your soul? It should be noted that these choices may not seem mutually exclusive. One may, perhaps, believe in an emergent mind, a brain that has a cool trick or two up its metaphoric sleeve, and that he or she has a soul. But the question is not whether those things exist, but which one of them is "you."

Part Two

NARROW AI AND ROBOTS

THE STORY OF JOHN HENRY

In the nineteenth-century folktale, John Henry is described as a "steel-driving man" whose job is to hammer steel spikes into rock. Historically, this was done to make holes into which explosives were inserted; these explosives cleared the way for new train track to be laid. John Henry was the best of them all. One day, a steam drill was invented to do that same job. John Henry wasn't about to be replaced by a machine, so he "told his captain, 'Well, a man's gotta act like a man. And before the steam drill beats me, I will die, hammer in my hand.'" So he challenged the steam drill to see which of them was better. It was a close contest, but John Henry prevailed. However, the work was so intense that John Henry collapsed and died of exhaustion on the spot immediately after the contest, hammer in his hand. People everywhere agreed, "John Henry died like a man."

6

Narrow AI

What exactly is artificial intelligence? This is not an easy question to wrap your brain around. To make an attempt, we need to go back to the origin of the term.

AI began as a real science around 1955. John McCarthy, a math professor at Dartmouth, decided to organize a project to explore the possibilities and limits of "artificial intelligence," a term he had coined the previous year. The goal was to "find how to make machines use language, form abstractions and concepts, solve [the] kinds of problems now reserved for humans, and improve themselves." To do this, he put together a group of four computer scientists who were already thinking about machines that could think: himself, Marvin Minsky, Nathaniel Rochester, and Claude Shannon, whom we met earlier when we discussed the first chess-playing computer program.

Their proposal then added a very, shall we say, "optimistic" prediction, especially given that it was 1955: "We think that a significant advance can be made in one or more of these problems if a carefully selected group of scientists work on it together for a summer."

McCarthy later regretted using the term "artificial intelligence" to describe this concept, feeling that it set the bar too high. He wished he had called it "computational intelligence" instead. Many people in the industry still agree and try to distance themselves from the term, although their motivation often seems to be a business one, to carve out a new area in AI that they can lay claim to pioneering.

I think the term "artificial intelligence" is great except for two problems: the word "artificial" and the word "intelligence." The word "artificial" is a problem in that there are two very different meanings of the word, and it is unclear which one is being used. Is it artificial in the sense that it is something that looks like intelligence but isn't, like the word "artificial" in "artificial turf"? Or should the word "artificial" be taken to mean it is true intelligence, but just artificial as opposed to natural or biological? The word "intelligence" is also a problem because there is no consensus definition of what intelligence is. The candidates cover an incredibly wide range of abilities.

But as the saying goes, "You have to dance with the date who brought you." Artificial intelligence is the term that got us here, and given that Chinese checkers is neither from China nor checkers, Arabic numbers were created in India, and koala bears aren't bears, using the term "artificial intelligence" is probably okay.

The broadest definition is that AI is technology that reacts to data or its environment, which means that a sprinkler system with a rain sensor is a kind of AI. A more strenuous definition is that it is a technology that learns from its environment. In this sense, a thermostat that learns the temperature at which you like your rooms is an AI, but the sprinkler system isn't.

But—and this is really important—there are two completely different things people mean today when they talk about artificial intelligence. There is "narrow AI" and there is "general AI." The kind of AI we have today is narrow AI, also known as weak AI. It is the only kind of AI we know how to build, and it is incredibly useful. Narrow AI is the ability for a computer to solve a specific kind of problem or perform a specific task. The other kind of AI is referred to by three different names: general AI, strong AI, or artificial general intelligence (AGI). Although the terms are interchangeable, I will use AGI from this point forward to refer to an artificial intelligence as smart and versatile as you or me. A Roomba vacuum cleaner, Siri, and a self-driving car are powered by narrow AI. A hypothetical robot that can unload the dishwasher would be powered by narrow AI. But if you wanted a robot MacGyver, that would require AGI, because MacGyver has to respond to situations that he has not previously considered. AGI does not currently exist, nor is there agreement on how to build it—or even if it is possible. That's what part four of this book is about. For now, we will focus solely on narrow AI, which despite having a wimpy name, is really amazing technology. Narrow AI is also not "easy AI," as the overwhelming majority of the money and sweat being spent on AI is on this type of artificial intelligence. From here on out, when I use the term AI, I am referring to narrow AI.

AI has developed to a point where it touches our lives several times a day. This has been a long time coming, for progress in AI has slowed to a crawl several times throughout the last few decades as investment in the technology dried up due to disappointment over its slow development. These periods of drought were known as "AI winters," and they lasted until some new technology or technique rekindled enthusiasm for AI, starting the cycle over again. AI has now proved itself in so many areas that it is unlikely that we will have another AI winter. This surge of progress has led the CEO of IBM, Ginni Rometty, to predict that by 2021 "cognitive AI will impact every decision made."

So how does AI work? Very broadly speaking, there are three different approaches to how to build AI. Let's say you want to make an AI that tells farmers when to plant their seeds. The first approach would be classic AI. We call this classic AI because in the earliest days of AI research, this is the approach that computer scientists thought would work the best. Classic AI involves thinking through all the factors (soil type, crop, rainfall, etc.) and building a model that takes those factors, weighs them accordingly, and makes a suggestion.

The second approach is called an expert system. An expert system is developed by taking a hundred of the best farmers and getting them to write down every rule about planting they know. You then arrange those rules in a way that someone can enter their relevant variables and the system will make a suggestion based on those rules.

The third approach is machine learning. Machine learning refers to a process by which you take all the data concerning when everyone planted and what the yields were and you get a computer to build rules that, in retrospect, would have maximized those yields. The tricky thing about machine learning, however, is that while its suggestions may work, they can be indecipherable to a human. The machine-learning AI may say, "Plant corn on March 12," and if you ask, "Why?" it may be hard to tease out an answer if there are many factors at work.

It is advances in this final area, machine learning, that have propelled AI forward recently. Large data sets, colloquially called "big data," combined with powerful computers and cunning algorithms have largely been responsible for the current renewed excitement about AI and its recent successes.

The speed at which AI advances is likely to increase. Chip designers talk about growing the speed of their products *faster* than Moore's law predicts. Quantum computers will potentially supercharge Moore's law, and they are not simply a thing of science fiction. Google's Hartmut Neven thinks that within ten years there will be only quantum computers doing our AI heavy lifting, with the machines we have today becoming as outdated as VCRs. Combine increased processing power with the unfathomable 2.5 billion gigabytes of data that is produced every day, which can be used to train AIs, and you have the ingredients for unimaginably fast progress.

While there are all kinds of variants to these three basic approaches, they either involve creating a model, asking experts, or studying data. None of them is the "right" way to implement AI, as some problems lend themselves to one approach over another. While all three of these methods are legitimate ways to make AI, an AGI might require a completely different approach, because it is trying to do a different thing. In other words, making an AGI that can solve problems that it has not been programmed to solve may be profoundly different from making an AI that can solve a narrowly defined problem. No one knows for sure, and experts are split on the question.

A natural housing for an AI is within a machine that gives it an ability to interact with the physical world—a robot. Some people, in fact, say this is essential, that AI cannot advance beyond a certain point unless the technology is embodied and can learn by interacting with the world. Our ability to build better robots is advancing at a steady, although somewhat plodding, pace as we develop new alloys, more efficient batteries, better sensors, and more efficient methods of locomotion. But what has really driven renewed excitement about the field of robotics is not so much those advances, but rather the prospect of those advances being combined with the supercharged artificial intelligence that we are building. That combination should give us AI robots that can physically exist in and interact with our physical world. That's our next stop.

7

Robots

I learned the story of the steel-driving man, John Henry, in song form at summer camp when I was ten. Even then I thought the whole story was crazy. Why didn't John Henry just get a job running the steam drill? He seems like he'd be a great candidate for the job, and it was bound to have paid better and been a whole bunch easier. Why would he prefer to spend his energy and vitality on such a job when he could operate the machine? And the idea that John Henry's death was somehow noble struck me, even as a child, as ridiculous. He threw his life away to try to prove an unprovable point, that human muscle power would always exceed machine power. And finally, John Henry must have used tools, such as the hammer he died holding. He didn't insist on hammering spikes with his fist. So why would he object to the steam-powered hammer?

It is a depressing tale, and to make matters worse, there is some evidence this legend is based on a true story.

Can you imagine a warehouse that stores bricks getting a forklift and someone who used to carry bricks on his back refusing to use it, challenging the forklift to a contest? Or a mathematician who would rather die with a slide rule in his hand than use an infernal calculator?

However, the story of John Henry does a pretty good job of summing up humanity's long and complex relationship with laborsaving devices.

The desire for robots to do work for us is far from new, and one does not have to look through ancient literature at great length to find a whole host of examples. In his *Politics*, Aristotle wrote in passing of the statues built by Daedalus, which moved around and had to be chained to the walls lest they wander off. He also mentioned the tripods of Hephaestus, which, according to Homer, went up and down Mount Olympus on their own. This same Hephaestus, in some accounts, built the mechanical bronze eagle that daily plucked the liver of Prometheus. Aristotle suggests in this work that slavery will end when men have invented contrivances to do all their work for them. The question of our day is whether technology will end the need for workers.

All through ancient myths from all around the world, artificial people abound. Sometimes they are fierce soldiers, such as the bronze Talos who defended Crete. Usually they are animated by sorcery instead of mechanisms, and their artificial bodies are imbued with will and set to a specific mission.

Once we entered the scientific era, the robots in popular culture ceased to be animated by magic and were increasingly powered through more modern means. In Mary Shelley's 1818 novel *Frankenstein; or, the Modern Prometheus*, the robot is made not of metal but of human body parts, and is brought to life using science.

As we approach more modern times, things get even more interesting. When Darwin's *On the Origin of Species by Means of Natural Selection* came out in 1859, a copy made its way to New Zealand and into the hands of a sheep farmer and writer named Samuel Butler, who devoured the text by candlelight. It struck him that if you applied Darwin's principles to machines, they would evolve to be conscious and to supplant us on this planet. In his 1863 letter to the editor of *The Press*, Christchurch, entitled "Darwin Among the Machines," he seems positively ecstatic about the possibility:

> *We refer to the question: What sort of creature man's next successor in the supremacy of the earth is likely to be. . . . It appears to us that we are ourselves creating our own successors. . . . In the course of ages we shall find ourselves the inferior race. Inferior in power, inferior in that moral quality of self-control, we shall look up to them as the acme of all that the best and wisest man can ever dare to aim at. No evil passions, no jealousy, no avarice, no impure desires will disturb the serene might of those glorious creatures.*

It is tempting to want to engage Butler as to the desirability of this outcome. One might begin by quoting Abraham Lincoln's observation that those "who have no vices have very few virtues." A machine with no avarice or impure desires likely has neither nobility nor compassion either.

In 1872 Butler published a novel called *Erewhon* in which he further developed his thoughts. The conclusion of the novel was that if one wanted to avoid the ascendency of the machines, there was only one recourse: outlaw them all, right down to the last gadget.

By the time the twentieth century rolled around, robots were completely creations of science. The word itself is derived from a Slavic word for slavery, and it was coined by the Czech author Karel Čapek in 1920 for his play *R.U.R.*, the title of which stands for "Rossum's Universal Robots." In the play, the character Harry Domin, who invented the robots, makes a prediction which sounds identical to those made by the techno-optimists of our time, a century later:

> *In ten years Rossum's Universal Robots will produce so much corn, so much cloth, so much everything, that things will be practically without price. There will be no poverty. All work will be done by living machines. Everybody will be free from worry and liberated from the degradation of labor. Everybody will live only to perfect himself.*

The robots in that play were manufactured using biomechanics, and they were designed to do whatever drudgery humans didn't want to. Their souls were not crushed by this work, because they were made without souls or feelings or passions of any kind. Eventually the humans stop doing any work and the robots do everything until they finally decide to take over and kill all the humans.

In most of their modern incarnations, robots serve either as the helpers of humans or their comic foils. From the replicants of *Blade Runner* to everyone's favorite protocol droid, C-3PO, robots are obviously still a mainstay of fiction, filling any number of roles. You might remember that the Jetsons had Rosie the Robot, a mechanical maid. One interesting thing to note is that that show was made in 1962 and set in 2062, meaning we are now closer to the time of its setting than the time of its creation. Still no sign of Rosie, though. That being said, the robots we do have are in some ways more amazing than Rosie. We have thousands of things we count on and enjoy that could not be made by humans, such as the CPU in your computer. Many of the jobs they do for us are simply ones that would go undone if the robots didn't do them, because they are beyond human ability.

Our desire for robots has long been to build them to do the three *D* kinds of jobs. These are jobs that are dirty, dangerous, and dull. You could add another four *D*s: disliked, demeaning, draining, and detestable. We want to give all these jobs to robots, who don't mind doing them.

Some even hope to give *all* jobs to robots. In this view of the future, the factories run themselves, and all the needs of life can be met with no human labor involved. In this world, so much wealth is created by technology that the concept of "working for a living" no longer has meaning, for the age-old link between work and survival has been broken.

In addition, many hope that we will build robots that will assist us in our everyday lives, to be our helpers. Toyota is investing a billion dollars to build robots to help the elderly in anticipation of the shortage of human caregivers brought on by longer lifespans and lower birth-rates. Sony is working on building robots that can connect emotionally with humans. These devices really could promote human dignity by empowering people with self-sufficiency. Others hope for disaster-relief robots that can be deployed in parts of the world that experience calamity.

What are our fears about robots? There are a few. The primary fear is that robots will compete with us in the job market and that we will lose to them. To date, robots have been an economic boon for humans, freeing us up to do ever more complicated and valuable jobs. But what happens at the end of that curve? What happens when a robot can do many or most jobs better than a person?

This fear is compounded by a basic characteristic of manufacturing. Over time, prices tend to go down as the quality of goods produced goes up. So robots will become better and cheaper, indefinitely. Surely, we fear, at some point robot labor will be both cheaper and better than human labor. Will the less skilled be locked out of the job market forever? If a fast-food robot costs $10,000, would a business rather pay a human a $15 minimum wage—or a $10 minimum wage, or *any* minimum wage? A massive displacement of this sort would represent a dramatic shift of economic power away from labor and toward those who own the robots.

Another widespread concern is the potential for a *WALL-E* future. In that world, we become permanently sedentary when we no longer have to work. On top of that, our brains atrophy as well, since the machines maintain themselves, as well as everything else.

Some fear that we will form emotional attachments to robots so strong that they will supplant bonds between humans. You don't necessarily need to go visit Grandma when she has an attentive robot helper. A robot companion may even be a suitable substitute for a spouse, and robot friends will probably be more reliable than the organic variety. But this is not universally a fear, as more than a few people hope this future will come about and that we will interact with robots in these ways.

And of course, the ultimate fear is the robot uprising. Although this is often spoken of with tongue firmly in cheek, there are several popularly imagined scenarios in which the robots themselves become adversarial, either through human malevolence, technical error, or an unexpected emergent behavior. It's a fact that militaries around the world are currently working to build better and better killer robots.

8

Technical Challenges

The combination of narrow AI and physical robots is more than the sum of its parts. When they come together, you get a seemingly intelligent agent that can act autonomously. This is where the hopes and the fears come into being. Roboticists are starting to build and deploy more and more robots with narrow AI, and when people see them for the first time, they are often not entirely sure what to make of them. Because the technology is new, we don't understand its limits, and therefore its implications. We don't know what these devices will ultimately be able to do.

Most of the hopes, fears, and questions concerning these technologies are rooted in what will happen with them in the future. This is because in spite of all of the great strides in both AI and robotics that we have made, glaring deficiencies in both are immediately obvious to anyone who interacts with these technologies daily. Let's explore just a few of these.

The first problem AI robots have is seeing. We can put a great camera in a robot, but that just gets us data. The robot has to sort it all out. When you look in your kitchen's pantry, you see a bunch of individual items. But a robot just sees pixels, a kaleidoscope of millions of specs of light and color. It doesn't know what a box is, or a shelf, or even an edge. It just sees an undifferentiated mass of dots. How do you even start to make sense of that? How do you get from those numbers to "That's a can of pork and beans"? It's quite hard. The number of things going on in your brain when you look in the pantry is complex in the extreme. A description of how your brain performs that minor miracle would require pages of technobabble about polygons and cones and layers. Your ability to, in an instant, identify the architectural style of a house, identify a dove in flight, tell the twins you know apart, or perform any of a hundred other similar tasks that you do effortlessly is the envy of AI programmers everywhere. I know some of those folks, and if they thought for a minute they could dissect your brain and find the answers to these challenges—well, I would suggest not accepting any cocktails mixed by them.

But let's say that you solve that problem, and the robot can recognize every item in the pantry. Robots can't even come close now, by the way. But let's imagine that they can. They still wouldn't be able to *understand* anything they "see." That's because AIs can't contextualize. If you were driving through town and saw a puppy in the road, a toddler walking toward it, and a grown woman frantically running toward the toddler, you wouldn't have trouble piecing that scene together. But to a computer, it would be just a bunch of pixels changing color. Really, again: it's just a bunch of ones and zeros. Think how easy it is for humans to figure out what is going on in a given photo: That one is a conga line. That one is people hiding for a surprise party. That's a prom photo taken by a parent. That's a piano recital, a school play, a christening, and so forth. Every one of those is easy for us because we have the cultural context to decipher it. Now in theory you *could* train a computer to do the same. Show the computer enough conga lines, and it will get really good at spotting conga lines. Better, in fact, than a human. That would be great if life consisted of nothing but still photos that need recognizing. But life is kinetic, it moves. Context is derived from the differences in a series of still images, and there are innumerable possible permutations of that. There are few training sets available with this kind of information. When you see your young neighbors, a husband and his very pregnant wife, hurry toward the car, with her holding her belly and him carrying an overnight bag and looking worried, you don't have any trouble figuring out what is going on. That's hard for a computer. But take it one step further. Say you didn't see them rush off, and it is now two days later. When you notice two newspapers in their yard and their car missing, you instantly have a good idea what happened. And you don't even have to think about it very hard, you just casually remark to your spouse, "Honey, I think the neighbors are having their baby." If your other neighbor has a son who is about to turn sixteen, and out of nowhere, he starts going door to door asking to mow yards, you might infer he wants to buy a car. Training a computer to make those intuitive leaps is quite difficult.

But, let's say you trained an AI to derive contextual clues from conga lines, pregnant wives, and sixteen-year-old boys. We can't even come close to this right now, but let's imagine that we can. Then we're still not much better off, because the AI can do only those things. We don't know how to teach it to do transfer learning. What is transfer learning? If I show you an item, let's say, a small statuette of a falcon about a foot tall, and then I show you a dozen photos and ask you to spot the falcon in them, you wouldn't have any trouble. Even if the falcon is half obscured by a tree or underwater or upside down or lying on its side or has peanut butter smeared on its head. I would wager that although you have never seen a falcon statuette with peanut butter smeared on its head before, you could still recognize the falcon. This is because humans can take a lifetime of experience of seeing things obscured by objects, underwater, or covered in substances like peanut butter, and apply that knowledge to a new task. That's transfer learning, and we don't understand how we do it, let alone how to teach a computer to do it.

But let's say we solved that problem. We haven't and likely won't any time soon. But let's say we have and the AI can take what it learns in one area and apply it elsewhere. We still aren't very far along, because the AI doesn't know how to improvise. Everyone, of every skill level, can improvise in a way far beyond any machine. If the door handle breaks off in your hand, you don't just stand there frozen, unable to fathom what to do in a universe you never contemplated, a universe in which doorknobs don't just turn, but break off in your hand. No, you try to figure out a way to get the door open. If you lock yourself out of your house, you figure out how to get in. If a gust of wind blows your umbrella away, you just go after it, without having ever been taught how. So even if the computer can see, understand what it sees, derive context from it, and use it in other spheres, it still isn't creative. We don't just passively perceive the world the way our hypothetical AI does, we react to it in ways that transcend transfer learning.

Beyond all this, we have four other senses to consider. Take hearing, for instance. Here we are in our great modern age and when I call my airline and tell its automated system my frequent-flier number, it gets it wrong about half the time. It turns out that recognizing speech is a pretty hard problem, even though in most languages there are only a few dozen possible letters or numbers a person might be saying. When you account for regional accents and the tonal differences in voices, it turns out your "H" and my "eight" are pretty close from a computer's perspective. Add a little static on the line, or someone vacuuming downstairs, or simply give a person a cold, and the problem gets even worse. Again, the data isn't the problem, but processing that data is. Separating out the vacuum cleaner sound is a really challenging problem. People have this knack of being able to tune out certain sounds, but we have no idea how we do it, and we certainly have no idea how to get an AI to do it.

And yet, even with all these limitations, our primitive AI already benefits our lives in a thousand ways. AI routes you through traffic, filters out spam, predicts the weather, suggests products for you to buy, and identifies credit card fraud. Smart cameras use it to identify faces, sports teams use it to improve their strategies, HR departments use it to select job candidates. AI reads handwriting, turns speech into text, and translates both of those into other languages. All these tasks are purely computational activities. Anyplace where we have lots of data, even if it is unordered raw data, is an area where AI can really excel. For instance, AI will soon take all the satellite data we have and find ancient cities for archaeologists to dig up, keep track of populations of wild animals, and monitor vegetation growth. Then it will take all the traffic data and use it to help us build smarter roads, time traffic lights more effectively, and reduce accidents. The list is endless. Kevin Kelly, the founding editor of *Wired* magazine, summed all this up when he tweeted, "The business plans of the next 10,000 startups are easy to forecast: Take X and add AI."

So why can an AI do all those things for us, but at the same time have all the limits we just described? It is because we're good at teaching AIs to do one thing at a time. If you want an AI to play chess or detect email spam, then you just teach it that simple thing. You don't ask it to generalize, contextualize, be creative, or anything else. Just teach it that one thing. And that is what we can do now pretty well. That is why it is called narrow AI.

In chapter 2 we discussed the virtues of the division of labor, one of the only "free lunches" in economic theory. It made civilization and prosperity possible. But it has an ironic consequence: the division of labor teaches that overall prosperity grows as someone hones his or her skills down to one focused task. This is not only true in manufacturing but also everywhere else. Say you are an attorney. You are a generalist. You make a good living. But how do you increase the rate you can charge? You specialize, perhaps in copyright law. Beyond that, you might specialize even further. And yet in doing so, you sow the seeds of your own occupational destruction. Ironically, the more you specialize, the easier you will be to replace with a machine. The more you increase your mastery of an ever-smaller domain, the easier it is to instantiate that knowledge in a computer program. A hunter-gatherer is much harder to build a computer replacement for than an X-ray technician, because the technician does just one narrow thing.

Ken Jennings, who was famously beaten on *Jeopardy!* by IBM's Wat-

son, explains that during that whole experience, the folks at IBM maintained a line graph that showed Watson's progress on its quest to the dot
labeled "Ken Jennings." Every week, Watson kept inching ever closer. In
his TED talk, Jennings explains how it all made him feel:

> *And I saw this line coming for me. And I realized, this is it. This is
> what it looks like when the future comes for you. It's not the Ter-
> minator's gunsight; it's a little line coming closer and closer to the
> thing you can do, the only thing that makes you special, the thing
> you're best at.*

Playing *Jeopardy!* is one narrow thing. Well, actually a few narrow
things. And that's why today's AI can master it.

So far, we have just been exploring the cognitive technical challenges
of AI robots. Let's look now at the equally formidable physical challenges.
The physical world is a difficult place for an AI robot unless it is perform-
ing purely repetitive motions inside a controlled environment, such as a
factory floor. In that environment, such robots really do well. There isn't
a human alive who can solder a billion transistors onto a computer chip
the size of a postage stamp, but robots can. Without robots making the
things that we use every day, we would have 1950s technology powering
a 1950s economy that produces a 1950s standard of living, at best. So if
you enjoy the prosperity and convenience of the modern world, thank
the next robot you come across.

Yet with all our technology, we don't know how to make a robot with
the physical prowess of a human three-year-old, let alone one better
than an adult human. Despite recent progress, robots, outside of factories, are still novelties, facing a laundry list of challenges, including
locomotion, sensing, and the manipulation of their environment. As the
roboticist Erico Guizzo sums up the state of the art:

> *Lots of people have been working on humanoid robots for decades,*
> *but the electric motors needed to drive a robot's legs and arms are*

*too big, heavy, and slow. Today's most advanced humanoid ro-
bots are still big hulking pieces of metal that are unsafe to operate
around people.*

The first challenge robots have is figuring out where they are. This is
both a sensing challenge and an AI challenge. Roboticists haven't really
even formulated best practices around how to do this. It varies by situation. Often a robot is tasked with making a map of where it is, then
keeping track of where it is on that map. This probably doesn't sound too
hard, because we do it effortlessly. But imagine the problem from the
point of view of a robot. A robot has been dropped into a room. It "sees" a
chair and a footstool. Since the chair and the footstool can be moved, the
robot can't use them as anchors. If it finds itself closer to the chair than it
was a minute ago, it doesn't know if the chair was moved, it was moved,
or both. As such, it has to constantly redraw its map. Building a map and
figuring out where you are on that map is called simultaneous localiza-
tion and mapping (SLAM). While it isn't an insurmountable problem by
a long shot, it's just one more thing that makes the job of being roboticist
a challenging one.

Then, of course, there is the challenge of powering the robot, especially if you want the mobility that batteries enable. We are far from
cracking this one. To offer just one example, in 2016, an enterprising
AI-enabled Russian robot named Promobot, which had been programmed to be autonomous, managed to escape the testing facility and
made it 164 feet before running out of power in the middle of the road,
causing a half-hour traffic jam. So much for the robot uprising.

Another big problem robots face is interacting with objects. While
robots can be much stronger physically than humans and operate in
more extreme environments, at present we are generally much more
agile across a wider range of tasks. Humans have a skeletal system with
two hundred bones overlaid with seven hundred or so muscles. It takes
six muscles simply to move something as small as an eyeball. Replicating that kind of flexibility in a machine is hard. To give you a sense of

the difficulty roboticists are facing here, consider the DARPA Robotics Challenge, which took place from 2012 to 2015. In 2015, the finals were held. Erik Sofge, writing for *Popular Science*, summed it up by saying that "the biggest and most well-funded international robotics competition in years was a failure."

The robots had to drive a car, navigate across rubble, use a doorknob, find a valve and shut it off, and so forth. They didn't have to do this all by themselves using AI. The challenge was whether they could perform these physical acts, not simply whether they could perform them without aid from humans. Additionally, the entrants knew what the robots would be asked to do. Even given that advantage, only a few of the twenty-four entrants finished a course that a drunken sailor on shore leave could do. This demonstrates the difficulty of building a mechanical human, let alone a mechanical superhuman.

For a human, the DARPA challenge doesn't seem all that hard. What could be easier than turning a doorknob and opening the door? A lot of things, it turns out. The robot has to identify the doorknob, navigate its hand to it, and squeeze it. Not too hard, not too soft. The robot needs to determine the friction of the knob itself. It then needs to turn it. Humans can tell pretty easily if the knob is turning or if their hand is sliding on the knob. That is very hard for a robot. Humans can tell when to stop turning because of resistance. The robot has to be trained to actually stop at some point before it breaks the mechanism. Then, holding the knob in the open position, the robot has to push. How hard? Well, that's extremely difficult to know ahead of time. How heavy is the door? Is it stuck?

And all of that is for naught if it turns out the door was one you pull, not push. The robot couldn't intuit that fact without being taught how. Imagine, then, how many things would be involved with teaching a robot to dig through rubble looking for earthquake survivors.

Touch is a big challenge for robots. The fact that the same human hands can be used to scratch a puppy's head and participate in a bar fight is a testament to their versatility, and to the challenges of replicating them. But even if you build a magnificent robotic finger, the robot still has to perceive what is going on at the end of it. Consider changing a baby, holding a kitten, or comforting a frightened child. Think of the nuance of the pressure and the timing of every touch that you do "without thinking." But a robot can't do *anything* without thinking. It has to dissect every action to the most minuscule detail. How would you program those actions? Reducing it down it to ones and zeros is obviously possible, but equally obviously difficult for a device that can only manipulate abstract symbols in memory.

One wrinkle with these sorts of perception problems is that we don't have the training data to teach the robots. Amazon has a huge database of "people who bought this also bought that" with which to train its recommendation engine. But we don't have all the tactile data of a million adults holding a million babies in a thousand situations. We could certainly collect the data by making a version of those CGI suits that people wear when making movies. Using upgraded sensors in the hands and fingers, we could get a thousand parents to wear them for a year to begin to collect that data. But no one is doing that right now.

The difficulty of robots interacting with the real world is evident in the attempt by Pieter Abbeel, a professor at the University of California, Berkeley, to teach a robot how to fold laundry.

The first challenge was perception: When you look at a pile of laundry, how do you tell where the shirt ends and the pants begin? Each load of laundry is completely unlike any other one. It is just a chaotic jumble. And don't get me started on the challenges of folding a fitted sheet, which according to many mathematicians is theoretically impossible. A person can use color, shading, and texture as clues, but what we do absentmindedly is incredibly hard for a robot. And what happens if your puppy happens to be napping in the clothes basket? To a robot, if it is in the basket, it's fair game for folding.

Undaunted, Abbeel's team worked on a simple version of the problem for years, and they finally got a robot to fold a towel in twenty minutes. Once they had that success to work from, they got the time down to under two minutes. Towels, however, are the easiest thing to identify and fold because they have nice right angles. With all that work, the robot still can't take an inside-out sock and right it. Abbeel sums it up: "Once you start working in robotics, you realize that things that kids learn to do up to age 10 . . . are actually the hardest things to get a robot to do."

Finally, when considering the capabilities of AI robots, both mental and physical, we must make one additional consideration. Even when we figure something out, what happens when things break? A PC is hardly a nascent technology, and yet I still have to reboot my machine at some point about every week. What happens when these robots malfunction? Of course, human beings malfunction in their own ways too. Airline pilots have heart attacks, pharmacists put the wrong pills in the bottle. But AI robot crashes may be different for a few reasons. First, because digital and mechanical systems are made to be duplicated, errors may be disseminated across countless instances. This would be analogous to all airline pilots having undiagnosed heart problems. Second, as mechanization becomes ever more deeply embedded, errors might be more systemic and harder to detect. A flaw in a system's internal clock, for instance, could lead to all kinds of catastrophic events. Finally, the kinds of systems we are talking about here will become ever more deeply interconnected and interdependent. The ripple effect of a small error might be quite large. Examples of this having already happened are numerous. Back in 1962, a NASA rocket that cost a bit under a billion dollars in today's money exploded in flight due to a single hyphen missing from deep within its code. In another example, a European rocket exploded in flight with $7 billion worth of loss because a 64-bit number was too large to convert to a 16-bit number, causing both a metaphoric and a literal crash. While expensive, these disasters were at least well contained. Imagine a similar problem affecting the self-driving car network, the power grid, or—gasp—your company's payroll system.

I point these issues out not to suggest that we should rethink our march toward a more mechanical future. Machines are, on the whole, more reliable than people in what they do. However, generally speaking, machine failures have a larger potential to cascade. Digital systems are generally more brittle than analog ones. Delete a word from *The Great Gatsby* and you still have a masterpiece. Delete a character from a compressed file and you have . . . alphabet soup. A missed hyphen won't make a human explode. We tend to be wrong lots of times but just by a little bit. Machines fail less often but to more catastrophic ends. So we should be mindful of how and where we implement technology.

9

Will Robots Take All Our Jobs?

The public discourse about robots is overwhelmingly about their impact on jobs, and therefore we will explore this topic in detail. The question at issue is this: Will automation, on net, eliminate more jobs than the economy will create, or will we remain close to full employment? An incalculable amount of analysis and opinion has been written on this question, and while it seems like a straightforward one, it is devilishly complex. That is why survey after survey of technologists, economists, and futurists show that they are almost evenly split on the answer to this "simple" question.

Why is there so little consensus? Isn't the answer just a simple calculation of jobs destroyed by technology offset by those created by it? Well, technically, yes, but that's a calculation easier described than done. The job calculation is complicated because we don't have a list of all of the jobs and the skills for each one. Even if we did, it constantly changes. We don't know what technology will be able to do over the next decade or so. We don't know the regulatory pressures and limits placed on the new technology, what will happen to the economy, the degree to which businesses will invest in new technology, what will happen to the minimum wage, how much the new technology will cost, how good it will be, and how many jobs it, in turn, will create. What specific breakthroughs in AI will occur? What new materials will scientists invent? What advances will be made in sensor technology?

And this is just the tip of the iceberg. We don't know how quickly society will accept robots in jobs held by humans, what shifts will happen in consumer demand, what will happen to the price of labor, what new trade agreements will be signed, which technologies will get locked up in the courts. We don't know how insurers will react to some robots, or how willing banks will be to loan money for purchasing them, or how truly human independent they will be. We don't know how many jobs their lowering of costs will create, how many jobs the robot-making industry will consist of, or how many jobs will be created that our present perspective prevents us from seeing. In short, our technology now moves forward faster than our ability to comprehend its implications.

This list goes on and on, and I offer it to show why the net effect of technology on jobs is not a back-of-the-envelope calculation. The margin of error introduced by any of these variables would be enough to throw the entire calculation off. And taken together, they present an insurmountable obstacle to forecasting this way.

So if we can't calculate the net job loss from robots directly, what is our path forward? We can view the problem differently, and realize that there are just three possible outcomes, which can be sorted by the assumptions that each relies on. Each of these three outcomes is possible. There are simply too many unknowns to draw definitive conclusions about what is going to happen. What if the cost of labor falls at the rate that we saw the cost of transistors fall over the past few decades? What if that underlying assumption of economics, scarcity, is diminished by technology? Or what if robots and AI are actually not all that different in impact from steam power or electricity, which definitely changed the economy but did not cause the unemployment rate to rise?

I suggest we proceed with a mind open to a world changing in unpredictable ways. But regardless of how it all unfolds, there are only three possible outcomes:

1. Robots and AI will take *all* the jobs. "All" should be taken to mean that machines can do every single thing better than humans and the only reason people hire a person over a machine is for purely emotional or nostalgic reasons. This view says the machines will paint better paintings, write better sitcoms, be better presidents . . . the whole ball of wax. As our rule of thumb, let's say "all" here is 90 percent of the jobs.

2. Robots and AI will take *some* of the jobs. This position believes that there will be a substantial net human job loss to machines, that the machines will destroy substantially more jobs than they will create. This position assumes the machines will take over most of the jobs in retail, service, delivery, and even a good deal of "professional" jobs like bookkeepers, doctors, and paralegals. But they don't take over the jobs in the arts, for instance. Nor can they do the jobs that require emotion or social skills, or that are too complex in the breadth of requirements (think CEO or installer of aftermarket car stereos). For the "some" position, let's assume there is a 20 percent level of unemployment that persists indefinitely. This is roughly the level of unemployment during the Great Depression in the United States. In this view, there simply aren't enough jobs for all the people, but the majority of people will still work.

3. Robots and AI will take *none* of the jobs. This position states that while jobs will be eliminated by machines, roughly the same number of jobs will be created, and that *everyone* remains employable. This position says you won't have a lack of work for any portion of society, even those with few job skills. The main idea behind this view is that it doesn't matter how good machines get, they simply cannot take all the jobs because there is an infinite supply

of them. Even if all the jobs of today were magically done by machines tomorrow, we would invent entirely new ways to make or do something that other people will pay for.

We'll examine each of these positions in detail.

One final note before we begin. Until this point, I have used the term "robot" in the sense in which most of us think of it, as a mechanical device that can operate autonomously. But from this point on, we need to broaden the term a bit. Robots, strictly speaking, don't necessarily need to be embodied or have locomotion. When considering the effects of automation on jobs, this is certainly the case. A machine replacing a beekeeper, tending hives and harvesting honey, is clearly a robot. However, a machine replacing a bookkeeper, keeping the accounting records in order, is also in this usage a robot as well. Even though that machine is a desktop computer and has no body, the end effect is the same. Both of these devices perform labor that was previously done by a human. Whether the machine is pushing around atoms or bits doesn't matter too much when we are talking about employment.

Possibility One: The Machines Take All the Jobs

Consider the foundational questions from earlier in the book. Possibility one, that the machines take all the jobs, will most likely appeal to those who believe that they themselves are machines. It will also likely resonate with those who believe that the "self" is a trick of the brain. In addition, this is a position that is most in alignment with monism, the belief that everything in the world is made of a single substance. Possibility one is based on a completely mechanistic view of the universe.

Let's start off by looking back the last couple of centuries through the lens of someone holding this position. They may describe it as such:

For 250 years, machines have been taking jobs from humans. At first, machines were able to do only the most basic, simple tasks. They took on the backbreaking work in the form of that marvel the steam engine. The things that steam could do had been such grinding toil for men and beasts that humanity welcomed the new machines with open arms. This was the start of the Industrial Revolution.

As technology improved our machines, they took on more complex, though still monotonous, tasks. Think sewing machines, cash registers, and combine harvesters. They were welcomed by the workers they freed from drudgery. And those same workers were pleased that they could now focus on higher-value work, leaving the repetitive labor to the machines.

But the march of progress never slows. The machines continued to advance, until suddenly, out of the blue, humans found themselves doing something they had never imagined: competing with machines for their own jobs. It seemed like a *Twilight Zone* episode gone bad, but Rod Serling didn't step out from behind a bush. This was real. Then came the computer.

As happened with mechanization, computers entered the scene as primitive devices. At first, they were simply glorified calculators, welcomed by scientists and mathematicians since they could do all of the tedious math problems. But they too began advancing in ability, doubling in power every two years. Before long, the bank tellers were facing off against the ATMs, while the stockbrokers battled against online trading websites. It was John Henry versus the steam drill all over again, only instead of competing with our manual labor, they were now providing cognitive labor as well.

Progress is relentless. Machines of all types will be improved upon indefinitely. You can see where this is heading: at some point soon, *technology will advance faster than the time it takes people to learn or invent new jobs.* As has happened in the past, the time between when a technological disruption creates a job and when another disruption destroys that job will get ever shorter, until the time will come when the machines can learn any new task better and faster than any human.

This isn't a guess. It is simply math.

Or, less abstractly, the position goes like this: "When you took the person who hauled bricks with a wheelbarrow and gave him a job hauling bricks with a forklift, the training required to make that transition was economically reasonable given the value of the newly trained worker. When our machines went from being super dumb to just plain dumb, we still had no problems mastering them. But the new jobs will require such extensive skills that it will not be economically viable to retrain workers to do them. And this all repeats, ad infinitum, progressively eating away at all meaningful human employment."

Most of the people holding this view believe that *all* human work will eventually be done by machines, and that the only economically profitable work left will be places where humans *subjectively* prefer hiring another human, even though the machine is *objectively* better. At most, this reasoning goes, you will be left with a few ballet dancers to entertain people who don't want to watch an anatomically perfect android dance *Swan Lake*, as well as a few craftsmen to make tchotchkes for nostalgia buffs who prefer anachronistic man-made items.

To many, this narrative is a frightening scenario. What would this world be like? And if machines are better at everything than us, either we are inferior to them in every conceivable way, or we are machines ourselves, and not particularly good ones. These fears are compounded by a tendency of the media to report on automation in such a way that projects intentionality, and even malice, onto the machines in headlines like "Is a Robot Gunning for Your Job?" or "Will a Computer Steal Your Job?" If a computer really stole your job, that would imply it snuck into the payroll system at night and deleted your name and added its own, chuckling to itself on its way out the door.

But putting aside all the emotional implications of this position, let's coolly consider the question of whether this viewpoint is true. Will we all, each of us, be replaceable by a machine at some time in the future? If possibility one is wrong, where does the logical breakdown occur? Let's look. There are nine assumptions underlying this argument.

ASSUMPTION 1: *Humans are machines.*

ASSUMPTION 2: *Since humans are machines, we can build a mechanical one.*

ASSUMPTION 3: *Mechanical humans would have the full range of our mental abilities, including creativity.*

The assumptions that humans are machines, and therefore it is possible to build a mechanical human, which would then possess the entirety of human mental ability, relate to the nature of our brains, which we addressed in our foundational questions. If you don't think you are a machine, then this argument falls apart before it even gets going. The proposition that we are machines is absolutely critical here.

Many people find the idea that we are machines to be unsettling and even a bit offensive. Others, however, fully embrace it. Marvin Minsky, who was an AI researcher for over half a century and truly one of the giants in the field, often referred to humans as "meat machines." He meant it literally. Ray Kurzweil longs for the day he can back up his "mind file" onto a computer, to be restored if he has an untimely death. Stephen Hawking said it plainly: "I regard the brain as a computer which will stop working when its components fail. There is no heaven or afterlife for broken-down computers; that is a fairy story for people afraid of the dark." The list goes on and on. To many, this is the inescapable conclusion of a reductionist view of the world.

But if we are machines, can we build one? And if we did build one, would it be creative, and have a mind and a will of its own? We certainly don't know how to build one, nor do we even understand how it is that we are creative and have a mind. We don't know where ideas come from, or how they are encoded in the brain. There is a broad range of human aptitudes that we don't understand at all. To be able to do everything a human can do, machines would need to acquire all of this. All these questions will come up again later in this book when we talk, in depth, about conscious computers.

ASSUMPTION 4: *This conscious machine would want to do our dirty work; and,*

ASSUMPTION 5: *Whether it is willing or not, we will compel it to, creating de facto mechanical slaves.*

Likewise, the assumptions a conscious machine would do our dirty work and that we would compel it to do so regardless of its will relate to what rights these mechanical beings may have. It could be that the minute you are able to build a robot assistant with true artificial intelligence, it might decide it would rather write poetry than iron your socks. Or maybe it would want to build *itself* a robot assistant to do the sock work.

ASSUMPTION 6: *It would be economically practical to build such mechanical humans.*

It is not certain that it would be economically practical to build mechanical brains. The Human Brain Project, a European endeavor to build an AGI modeled on the structure of the human brain, has already spent a billion dollars toward this effort, and as of this writing, *Scientific American* reports that the project is currently "in disarray." The brain is, all agree, the most complex item we know of in the universe. A mechanical one may cost dramatically more than an iPad. Or it may not, for our experience is that prices of electronic devices fall rapidly, and that might well happen with respect to electronic brains. We just don't know. Some experts in AI believe that the computational power an AGI would need is quite small—less than what your smartphone has—if only we had the software worked out. So a machine with AGI may not be particularly hard to build and, being mostly software, might cost just a few cents. That would sure be humbling. Or perhaps AGIs might be exceedingly expensive, but we would need to build only a few that we all share.

ASSUMPTION 7: *Machines would become so inexpensive or efficient that they would be cheaper to deploy than human labor.*

ASSUMPTION 8: *The programming cost to teach the machine a new skill plus the cost of running the machine will always be less than the labor costs of paying a human to do it.*

The assumptions machines will become so inexpensive or efficient that they'll be cheaper than human labor, and that the cost to program and run these machines will be less than the cost of paying a human to do the machine's job, both relate to the economic viability of using such machines to do low-cost labor. For all jobs to be lost, the costs of building, programming, and operating such machines will need to be less than the cost of employing a human in those jobs. It might be that the cost of developing machines to do obscure tasks will fall rapidly over time, or it might not. Andrew Ng, one of the industry's intellectual heavyweights, thinks this will happen:

> *There will be AI that gradually learns to do everything we do. And when a machine can do almost everything better than almost everyone, our social structure will begin to unravel. And that's something we need to prepare for.*

ASSUMPTION 9: *Humans lack the ability to find other tasks that machines cannot do.*

The final assumption is that humans lack the ability to discover new jobs that can't be done by machines. For this to be the case, the minute we invent a new job, a machine can be deployed to do it faster and cheaper than a human.

If all nine of these assumptions play out, then virtually all paid employment to humans will vanish and most institutions in society will have to be rethought. All manners of utopias and dystopias pop to mind when imagining life in such a world. If any one of these assumptions is not true, then the argument collapses. That being said, many regard all these assumptions as obvious if you begin with the idea that we are machines, that computers will continue to improve, and that the cost of building technology will continue to fall. Those three ideas taken together mean that sooner or later, the machines will surpass us in everything.

Possibility Two: The Machines Take Some of the Jobs

Possibility two is that the machines take some of the jobs, which results in long-term unemployment. If we recall our earlier foundational questions, this possibility might resonate with those who regard themselves as either animals or humans, as well as the dualists, who acknowledge that in addition to a physical world, there is a mental world or a spiritual world. It will also appeal to those who see their "self" as either an emergent property or their soul. All of these various beliefs have at their core the idea that there are substantial differences between humans and machines, and thus there are limits to what machines can do. But these beliefs also allow for a good deal of overlap between the capabilities of humans and machines.

The logic of the "some" position is straightforward, and is a narrative that will be familiar to many readers. It goes like this:

> Of course, job loss due to technology has existed in the past, but it was always offset by technologies creating new jobs. Now, however, times are different. How? First, the rate of innovation

is much faster, so jobs are eliminated much more quickly. Second, in the past, automation replaced only physical labor, but now we are dealing with the widespread automation of cognitive labor, which will destroy many previously untouched industries. And finally, the cost of automation technology is falling rapidly, meaning that a robot that costs $1 million in 2020 may cost only $1,000 in 2030.

The net of all of this is that we are about to experience job loss in a much more dramatic way than ever before. And it has been shown that the lower-paying a job is, the more prone it is to automation, generally speaking. This means that the lowest-skilled workers will lose their jobs first, and collectively, this group will be competing for fewer and fewer jobs.

While it is true that technology will create some new jobs, they will be relatively few in number, and they will require extensive education and training. Sure, you may open a factory to make order-taking robot kiosks for fast-food restaurants and create a few new high-wage jobs, but that factory can manufacture many thousands of those robots, destroying untold numbers of order-taking jobs. If technology is destroying vast numbers of low-skilled jobs but is creating only a few new high-skill jobs, we will be left with a shortage of low-skill jobs and a large number of permanently unemployed low-skilled workers. Hence, we are in a permanent Great Depression.

There are jobs that computers won't be doing, or at least won't for centuries. Priest, plumber, and policeman come to mind. There are real limits to what the machines can do. However, those limits are well above the requirements of many, many existing jobs.

Once you accept the idea that automation will do more and more low-skill jobs, you cannot escape the fact that this will result in too many low-skilled workers and too few low-skill jobs.

What do we say to this line of narrative? Let's explore the five assumptions behind it.

ASSUMPTION 1: *Machines and technology cause a net loss of jobs.*

The accusation that technology is a net job destroyer has been argued for a long time. In the 1580s, William Lee invented the stocking frame knitting machine. He pulled a few strings and arranged to give a demonstration of his device to Queen Elizabeth in the hopes of obtaining a royal patent. The queen thought the device cunning but remonstrated Lee, saying, "Consider thou what the invention could do to my poor subjects. It would assuredly bring to them ruin by depriving them of employment, thus making them beggars." Lee, in fact, had to leave England because of the anger of the hosiers.

As rapid manufacturing advances swept through the world in the next few centuries, every invention was greeted with anger and hostility from labor. French textile workers resisted the automated looms by throwing their wooden shoes into the machinery. In England, participants in the Swing Riots resisted automatic threshing machines by smashing them. Boatmen destroyed the first attempts at a steam engine, which they felt would put them out of work. So overwhelming was the protest against ribbon looms in Germany that they were ordered burned by the government. When the fly shuttle was invented to make weaving easier, its creator, John Kay, was attacked by a crowd. James Hargreaves, who created another breakthrough in textiles called the spinning jenny, saw his creation burned by yet another mob in England. John Heathcoat, who created technology to make the creation of lace more efficient, saw his entire factory and its equipment torched in broad daylight.

In 1811, this hostility toward automation coalesced into the Luddite movement, which consisted of a group of people violently opposed to technology that replaced skilled laborers. Taking as its namesake Ned Ludd, a youth who is said to have smashed two stocking looms, it attracted hundreds of proponents who roamed the countryside, burning factories and, in some places, killing the owners of the machinery.

One story from this time may show us a path forward. It happened on November 29, 1814. It was on this date that the *Times* of London was first printed using steam-powered printing presses. The pressmen vowed vengeance on the machine's inventor and destruction to the machine itself. However, they were told that if they refrained from violence, they would all be kept on at full pay until similar jobs could be found for them elsewhere. This seemed fair to the pressmen, and the march of progress continued.

These violent and destructive reactions are understandable. Most laborers of these times lived at or below the poverty line. Men and women spent their lives mastering a trade, and then just barely eked out a living. Economic opportunities were scant. The idea of competing with machinery and losing one's livelihood in the bargain seemed too horrific to contemplate. These workers must have thought that the wealthy factory owners had such an unquenchable desire for money that they would have eliminated the job of every worker if technology had allowed it. This may even have been true.

These extreme reactions are not in response to technology in general, but to laborsaving technology in specific. You don't read about the Great Air Conditioner Riots of '49 in history class for a simple reason: people don't riot over technological inventions that don't replace human labor.

This tendency to be hesitant of the consequences of new technology lives into the modern era as well. The Kodak camera was predicted to destroy art. Electricity was so widely feared that in 1891, President Benjamin Harrison had servants turn the lights on and off in the White House, because neither he nor his family would touch the switches. When car radios became widespread in the 1930s, it was argued that people would fiddle with them while driving, or even if they didn't, they would still get distracted and get into wrecks. Present concerns about GMOs, whether merited or not, are manifestations of this innate caution.

In addition to workers' practical concerns about automation, there are ideological ones as well. Karl Marx voiced his belief that machines were at odds with workers by saying, "The instrument of labour, when it takes the form of a machine, immediately becomes a competitor of the workman himself.... History discloses no tragedy more horrible than the gradual extinction of the English hand-loom weavers." The idea of people competing with machines at the same activity just seems wrong to some.

If indeed laborsaving technology does result in a net loss of jobs, we are in for some trouble. AI makes new breakthroughs by the day. More patents in technology are awarded each year than in the prior one. In the United States, almost a thousand are issued every day. Additionally, each of the last few years has set a new record in the number of robots sold worldwide. Costs are down a quarter over the last decade, and they are expected to fall another quarter in the next decade, and all the while, robots are getting better in quality. There is no reason to think that any of those trends will change. Amazon already has fifteen thousand robots working in warehouses, gathering products and preparing them to be shipped. The number of robots at Amazon isn't likely to decline.

There is a certain simple logic to this concern about net job loss, and it goes like this: There are only so many jobs in the world. And if you give a job to a robot, then that's one less job for a human being.

But the interesting thing about this line of thought is how obviously untrue it is. Most jobs that have ever existed have been displaced by technology. In no particular order, I offer some examples: Most stable-boys lost their jobs when the car was invented. Virtually every candle-maker got laid off when the kerosene lamp was fired up. In bowling, not that long ago, there were boys who stood the fallen pins back up, but they were replaced by, get this, robots. There were elevator opera-tors until someone came along and invented that job-eating technol-ogy the button. At one time, there were boys who delivered telegrams, and then somebody had to upset that applecart and go and invent the telephone, putting all of them out of work. At one time, you could count on streetlamp lighters to come out at dusk and do their part to battle darkness, until someone went and invented electric lighting. There once was an entire industry of people who delivered ice, until the advent of the electric refrigerator, which sent them all to the unemployment line. There used to be people who swept the streets clean with push brooms. Then they invented a vehicle to do it.

What happened to all these people who were displaced by technology? After all, you never hear stories of roving gangs of elevator operators unable to find work. The answer is obvious. All those people who saw their jobs eliminated went on to do different things. I don't mean to minimize how difficult these transitions are, but instead I point this out to highlight how extraordinary human beings are. Animals in nature can do only one set of things. Narrow AI, as I indicated earlier, can do only one thing. But our versatility borders on the infinite. The greatest underutilized resource in all of the world is human potential. And the more technology we can employ in our pursuits, the more we can do, and thus, generally speaking, the higher our wages will be.

Certainly we have seen technology reduce the need for workers in particular sectors. In just the twentieth century, agriculture went from being 40 percent of our jobs to 2 percent. And in just the back half of the twentieth century, manufacturing jobs went from 30 percent of our economy to 10 percent. In that century, we saw entire new professions created and then destroyed. The churn of jobs almost boggles the mind. Just mentally compare the workforce of 1900 with the workforce of 2000. I calculate that the half-life of a job in a modern economy is about fifty or so years. From 1900 to 1950, probably half the jobs vanished, mostly in farming. From 1950 to 2000, another half, many of which were manufacturing. And most, if not all, of this disruption was caused by technology. To believe that technology is a net destroyer of jobs, one must explain the fact that *all* this disruption happened during a period of full employment, rising gross national product, and rising wages. (The Great Depression, a decade that defies the trend, was not caused by technology, but rather by macroeconomic forces.)

But was the twentieth century in the United States anomalous? If we look beyond our own shores, one way to tackle the question is to compare places with lots of robots to places without them. If robots do replace people, then we should see a decline in manufacturing jobs in places where more robots are put in use in manufacturing. But as Mark Muro and Scott Andes of the Brookings Institution write in the *Harvard Business Review*:

> *[There is] . . . no visible relationship between the use of robots and the change in manufacturing employment. Despite the installation of far more robots between 1993 and 2007, Germany lost just 19% of its manufacturing jobs between 1996 and 2012 compared to a 33% drop in the U.S.*

They then go on to point out that other countries, such as Italy, South Korea, and France, lost a smaller percentage of manufacturing jobs than did the United States, even though they deployed more robots than the United States did. On the other end of the spectrum, other countries including the United Kingdom and Australia invested less than the United States in robots and had even bigger declines in manufacturing employment. The idea that robots would create manufacturing jobs and not destroy them is not as surprising as it might seem. When you can introduce efficiencies in an industry, you lower costs or increase quality. Lower costs and increased quality invite higher production, which creates jobs.

The truth is that technology mostly *augments* workers, not replaces them. The MIT economics professor David Autor argues that technology can't really do the vast majority of jobs.

> *Tasks that cannot be substituted by computerization are generally complemented by it. . . . Most work processes draw upon a multifaceted set of inputs: labor and capital; brains and brawn; creativity and rote repetition; technical mastery and intuitive judgment; perspiration and inspiration; adherence to rules and judicious application of discretion.*

He goes on to maintain that using technology to automate some part of a job almost always makes the tasks that the machine cannot do more valuable, because with technology, the value of the entire job goes up.

Lawrence Katz, a Harvard economist, echoes this view and maintains that history doesn't show any sort of net job loss due to technology. He told the *MIT Technology Review* that "we never have run out of jobs. There is no long-term trend of eliminating work for people. Over the long term, employment rates are fairly stable. People have always been able to create new jobs. People come up with new things to do."

Of course, in a few cases, new technology directly eliminates jobs. But history shows it creates offsetting ones as well. Sometimes this connection is hard to see. At the airport in Dusseldorf, robotic valet parking is now offered. You push a button and a giant machine lifts your car up and tucks it away in a vertical garage. There are new jobs—but they're not at the airport. They are at a factory making parking robots, employing people at wages presumably higher than a valet job. So this technology created new high-paying jobs.

What do we say about the kinds of jobs that the machines take, such as the valet job? Consider any job that can, in theory, be done by a machine. If you make a human being do that job, then that is literally "dehumanizing." There's nothing about that job that takes advantage of that person's being a human being. It's the worst possible work you can give anybody, to say, "You are just a stand-in until we build a machine to do this." The word "boredom" was invented only once we had factories. It was first used, in fact, in a Dickens novel, *Bleak House*. Factories are full of jobs that dehumanize people. That's why Dickens had to create a synonym for the result of unendingly repetitive and tedious work. Robots don't get bored, and frankly, humans deserve better. Clearly there are at least a billion people in the world who would love to have any job, even those jobs. The bottom billion live so close to starvation that any work is welcome. This much is obvious. But our long-term goal as a species should be to build machines to do those jobs so that people can do jobs that only people can do.

So the idea that technology is eliminating the need for workers seems a hard one to make. To get there, the "this time is different" assumption will need to be ironclad. We will get to that in a moment.

ASSUMPTION 2: *Too many jobs will be destroyed too quickly.*

The "jobs will be destroyed too quickly" argument is an old one as well. In 1930, the economist John Maynard Keynes voiced it by saying, "We are being afflicted with a new disease . . . technological unemployment. This means unemployment due to our discovery of means of economising the use of labour outrunning the pace at which we can find new uses for labour."

In 1978, *New Scientist* repeated the concern:

The relationship between technology and employment opportunities most commonly considered and discussed is, of course, the tendency for technology to be labour-saving and thus eliminate employment opportunities—if not actual jobs.

In 1995, the refrain was still the name. David F. Noble wrote in *Progress Without People*:

Computer-aided manufacturing, robotics, computer inventories, automated switchboards and tellers, telecommunication technologies— all have been used to displace and replace people, to enable employers to reduce labour costs, contract-out, relocate operations.

But is it true now? Will new technology destroy the current jobs too quickly?

A number of studies have tried to answer this question directly. One of the very finest and certainly the most quoted was published in 2013 by Carl Benedikt Frey and Michael A. Osborne, both of Oxford University. The report, titled *The Future of Employment*, is seventy-two pages long, but what has been referenced most frequently in the media is a single ten-word phrase: "about 47 percent of total US employment is at risk." Hey, who needs more than that? It made for juicy and salacious headlines, to be sure. It seemed as if every news source screamed a variant of "Half of US Jobs Will Be Taken by Computers in Twenty Years."

If we really are going to lose half our jobs in twenty years, well, then the *New York Times* should dust off the giant type it used back in 1969 when it printed "MEN WALK ON MOON" and report the story on the front page with equal emphasis. But that is not actually what Frey and Osborne wrote. Toward the end of the report, they provide a four-hundred-word description of some of the limitations of the study's methodology. They state that "we make no attempt to estimate how many jobs will actually be automated. The actual extent and pace of computerisation will depend on several additional factors which were left unaccounted for."

So what's with the 47 percent figure? What they said is that *some tasks* within 47 percent of jobs will be automated. Well, there is nothing terribly shocking about that at all. Pretty much every job there is has had tasks within it automated. But the job remains. It is just different.

For instance, Frey and Osborne give the following jobs a 65 percent or better chance of being computerized: social science research assistants, atmospheric and space scientists, and pharmacy aides. So what does this mean? Social science professors will no longer have research assistants? Of course they will. They will just do different things, because much of what they do today will be automated. There won't be any more space scientists? Pharmacists will no longer have anyone helping them?

Frey and Osborne say that the tasks of a barber have an 80 percent chance of being taken over by AI or robots. In their category of jobs with a 90 percent or higher chance of certain tasks being computerized are tour guides and carpenters' helpers.

The disconnect is clear: some of what a carpenter's helper does will get automated, but the carpenter helper job won't vanish; it will morph, as almost everyone else's job will, from architect to zoologist. Sure, your iPhone can be a tour guide, but that won't make tour guides vanish.

Anyone who took the time to read past the introduction to *The Future of Employment* saw this. And to be clear, Frey and Osborne were very up-front. They stated, in scholar-speak, the following:

> *We do not capture any within-occupation variation resulting from the computerisation of tasks that simply free-up time for human labour to perform other tasks.*

In response to the Frey and Osborne paper, the Organisation for Economic Co-operation and Development (OECD), an intergovernmental economic organization made up of nations committed to free markets and democracy, released a report in 2016 that directly counters it. In this report, entitled *The Risk of Automation for Jobs in OECD Countries*, the authors apply a "whole job" methodology and come up with the percent of jobs potentially lost to computerization as 9 percent. That is pretty normal churn for the economy.

At the end of 2015, McKinsey & Company published a report entitled *Four Fundamentals of Workplace Automation* that came to similar conclusions as the OECD. But again, it had a number too provocative for the media to resist sensationalizing. The report said, "The bottom line is that 45 percent of work activities could be automated using already demonstrated technology," which was predictably reported as variants of "45% of Jobs to Be Eliminated with Existing Technology." Often overlooked was the fuller explanation of the report's conclusion:

> *Our results to date suggest, first and foremost, that a focus on occupations is misleading. Very few occupations will be automated in their entirety in the near or medium term. Rather, certain activities are more likely to be automated, requiring entire business processes to be transformed, and jobs performed by people to be redefined, much like the bank teller's job was redefined with the advent of ATMs.*

The "47 percent [or 45 percent] of jobs will vanish" interpretation doesn't even come close to passing the sniff test. Humans, even ones with little or no professional training, have incredible skills we hardly ever think about. Let's look closely at two of the jobs at the very top of Frey and Osborne's list: short-order cook and waiter. Both have 94 percent chance of being computerized.

Imagine you own a pizza restaurant that employs one cook and one waiter. A fast-talking door-to-door robot salesman manages to sell you two robots: one designed to make pizzas and one designed to take orders and deliver pizzas to tables. All you have to do is preload the food containers with the appropriate ingredients and head off to Bermuda. The robot waiter, who understands twenty languages, takes orders with amazing accuracy, and flawlessly handles special requests like "I want half this, half that" and "light on the sauce." The orders are sent to the pizza robot, who makes the pizza with speed and consistency.

Let's check in on these two robots on their first day of work and see how things are going:

- A patron spills his drink. The robots haven't been taught to clean up spills, since this is a surprisingly complicated task. The programmers knew this could happen, but the permutations of what could be spilled and where were too hard to deal with. They promised to include it in a future release, and in the meantime, to program the robot to show the customers where the cleaning supplies are kept.
- A little dog, one of those yip-yips, comes yipping in and the waiter robot trips and falls down. Having no mechanism to right itself, it invokes the "I have fallen and cannot get up" protocol, which repeats that phrase over and over with an escalating tone of desperation until someone helps it up. When asked about this problem, the programmers reply, snappishly, that "it's on the list."
- Maggots get in the shredded cheese. Maggoty pizza is served to the patrons. All the robot is trained to do with customers un-

happy with their orders is to remake their pizzas. More maggots. The robots don't even know what maggots are.

- A well-meaning pair of Boy Scouts pop in to ask if the pipe jutting out of the roof should be emitting smoke. They say they hadn't noticed it before. Should it be? How would the robot know?

- A not-well-meaning pair of boys come in and order a "pizza with no crust" to see if the robots would try to make it and ruin the oven. After that, they order a pizza with double crust and another one with twenty times the normal amount of sauce. Given that they are both wearing Richard Nixon masks, the usual protocol of taking photographs of troublesome patrons doesn't work and results only in a franchise-wide ban of Richard Nixon at affiliated restaurants.

- A patron begins choking on a pepperoni. Thinking he must be trying to order something, the robot keeps asking him to restate his request. The patron ends up dying right there at his table. After seeing no motion from him for half an hour, the robot repeatedly runs its "Sleeping Patron" protocol, which involves poking the customer and saying, "Excuse me, sir, please wake up" repeatedly.

- The fire marshal shows up, seeing the odd smoke from the pipe in the roof, which he hadn't noticed before. Upon discovering maggot-infested pizza and a dead patron being repeatedly poked by a robot, he shuts the whole place down. Meanwhile, you haven't even boarded your flight to Bermuda.

This scenario is, of course, just the beginning. The range of things the robot waiter and cook can't do is enough to provide sitcom material for ten seasons, with a couple of Christmas specials thrown in. The point is that those who think so-called low-skilled humans are easy targets for robot replacement haven't fully realized what a magnificently versatile thing any human being is and how our most advanced electronics are little more than glorified toaster ovens.

While it is clear that we will see ever-faster technological advances, it is unlikely that they will be different enough in nature to buck our two-hundred-year run of plenty of jobs and rising wages. In one sense, no technology really compares to mechanization, electricity, or steam engines in its impact on labor. And those were a huge win for both workers and the overall economy, even though they were incredibly disruptive.

ASSUMPTION 3: *Not enough new jobs will be created quickly enough.*

The "we won't make new jobs fast enough" argument, you won't be surprised to hear, has been around for a while too. In 1961, *Time* magazine printed, "What worries many job experts more is that automation may prevent the economy from creating enough new jobs. . . . Today's new industries have comparatively few jobs for the unskilled or semiskilled, just the class of workers whose jobs are being eliminated by automation."

Is this a valid concern today? Will new jobs be slow in coming? I suspect not.

In 2016, the World Economic Forum in Davos, Switzerland, published a briefing paper that stated:

> *In many industries and countries, the most in-demand occupations or specialties did not exist 10 or even five years ago, and the pace of change is set to accelerate. By one popular estimate, 65% of children entering primary school today will ultimately end up working in completely new job types that don't yet exist.*

Think about that. By that estimate, two-thirds of kids entering school aren't going to exit as doctors, lawyers, architects, or *any* other profession that has a name today. They will be flummerflaves and wabberwocks and chort wranglers.

The two-thirds number may be too high. There is no way to know. But who would have ever thought that when you connected a bunch of computers together and allowed them to communicate using common protocols like http, it would create *trillions* of dollars of wealth in the form of Google, Facebook, Twitter, Alibaba, Amazon, Baidu, eBay, Etsy, and all the rest? Who could have guessed that this seemingly simple act would transform the world and everything in it?

MIT's David Autor has this to say about those who believe we aren't going to come up with enough new jobs, thereby rendering people obsolete:

> *These self-proclaimed oracles are in effect saying, "If I can't think of what people will do for work in the future, then you, me and our kids aren't going to think of it either." I don't have the guts to take that bet against human ingenuity.*

The legendary entrepreneur and technologist Marc Andreessen agrees:

> *I do not believe robots will eat all the jobs. Here is why. For a start, robots and artificial intelligence are not nearly as powerful and sophisticated as some people fear. . . . And even when robots and artificial intelligence are far more powerful, decades from now, there will still be many things people can do that they cannot. . . . Just as most of us have jobs that were not even invented 100 years ago, the same will be true 100 years from now. We have no idea what those jobs will be, but I am certain there will be an enormous number of them.*

We have this sense that more jobs are being destroyed than created because the destruction of jobs is more obvious and easier to see. When Foxconn replaces sixty thousand human workers with robots in one factory, you can bet it makes the front page, and the editorial page as well. When Adidas builds a plant to make shoes solely with robots, the story goes viral. However, the real story is the second-order effects that these events cause. For instance, I often hear Google Translate touted as a prime example of a technology putting people out of work.

It used to be, the reasoning goes, you needed a person to translate a document. Now you don't. Case closed. However, that isn't the whole story, is it? What really happens is that this powerful technology enables cross-language business transactions to occur more seamlessly. More translators are needed because there are many more contracts and other documents that need the additional precision of a human translator. Additionally, the human translators will expand what they do, such as helping businesses understand cultural differences with the people they are dealing with. All of this is work that would simply not have existed without Google Translate. That's why by 2024, the US Bureau of Labor Statistics projects a 29 percent employment growth for interpreters and translators.

ASSUMPTION 4: *Low-skilled workers will be the first to go.*
ASSUMPTION 5: *There won't be enough jobs for these workers in the future.*

The assumptions that low-skilled workers will be the first to go and that there won't be enough jobs for them undoubtedly have some truth to them, but they require some qualification. Generally speaking, when scoring jobs for how likely they are to be replaced by automation, the lower the wage a job pays, the higher the chance it will be automated. The inference usually drawn from this phenomenon is that a low-wage job is a low-skill job.

This is not always the case. From a robot's point of view, which of these jobs requires more skill: a waiter or a highly trained radiologist who interprets CT scans? A waiter, hands down. It requires hundreds of skills, from spotting rancid meat to cleaning up baby vomit. But because we take all those things for granted, we don't think they are all that hard. To a robot, the radiologist job is by comparison a cakewalk. It is just data in, probabilities out.

This phenomenon is so well documented that it has a name, the Moravec paradox. Hans Moravec was among those who noted that it is easier to do hard, brainy things with computers than "easy" things. It is easier to get a computer to beat a grandmaster at chess than it is to get one to tell the difference between a photo of a dog and one of a cat.

Waiters' jobs pay less than radiologists' jobs not because they require fewer skills, but because the skills needed to be a waiter are widely available, whereas comparatively few people have the uncommon ability to interpret CT scans.

What this means is that the effects of automation are not going to be overwhelmingly borne by low-wage earners. Order takers at fast-food places may be replaced by machines, but the people who clean up the restaurant at night won't be. The jobs that automation affects will be spread throughout the wage spectrum.

All that being said, there is a widespread concern that automation is destroying jobs at the "bottom" and creating new jobs at the "top." Automation, this logic goes, may be making new jobs at the top, like geneticist, but is destroying jobs at the bottom like warehouse worker. Doesn't this situation lead to a giant impoverished underclass locked out of gainful employment?

Often, the analysis you hear goes along these lines: "The new jobs are too complex for less-skilled workers. For instance, if a new robot replaces a warehouse worker, tomorrow the world will need one less warehouse worker. Even if the world also happened to need an additional geneticist, what are you doing to do? Will the warehouse worker have the time, money, and aptitude to train for the geneticist's job?"

No. The warehouse worker doesn't become the geneticist. What actually happens is this: A college biology professor becomes the new geneticist; a high-school biology teacher takes the college job; a substitute elementary teacher takes the high school job; and the unemployed warehouse worker becomes a substitute teacher. This is story of progress. When a new job is created at the top, everyone gets a promotion. The question is not "Can a warehouse worker become a geneticist" but "Can everyone do a job a little harder than the one they currently do?" If the answer to that is yes, which I emphatically believe, then we want all new jobs to be created at the top, so that everyone gets a chance to move up a rung on the ladder of success.

Possibility Three: The Machines Take None of the Jobs

Possibility three is that the machines, on net, take none of the jobs and that we will remain at essentially full employment. Recalling our earlier foundational questions, those who identify with possibility three have many of the same assumptions as those who identify with possibility two. Dualists can readily identify with this view, as can those who see the "self" as either an emergent property or the soul, something that would be difficult to reproduce mechanically. Possibility three is based on the idea that humans are something different from animals and machines, and therefore they have skills that neither can duplicate.

The narrative of the third position goes like this:

Machines simply cannot take jobs for three reasons:

> First, there exists a range of jobs that machines will not be able to do, jobs that we won't want them to do, and jobs that it isn't economical for them to do. Most jobs require dozens of skills like creativity and empathy that computers will never have, or at least not for centuries. While science fiction robots often have these skills, no one in his or her right mind looks at his or her smartphone and concludes that in a few years, it will be as intelligent as a human, and thus be able do any job a human can, from motivational speaker to ballerina. Computers have no agency, no presence. There is "no one at home." They run their ever-more-complicated programs and do wondrous things, but they are not the same type of thing as a human.
>
> Second, for the sake of argument, let's say the first statement is wrong, and that computers can, in fact, do 80 percent of all of

the jobs out there. Even then, there still won't be any unemployment. There are an unlimited number of jobs, because any time
people figure out how to sell a service they offer or a product they
make, they just created a job. Therefore, technology actually empowers people to create more and better jobs.

The misconception that there is a fixed pool of jobs that the robots are going to pick off one by one has a name, the Lump of Labor
Fallacy. But in reality, there will be unlimited jobs indefinitely, for
they are created by the human mind, not by an outside force.

Third, even if robots did all the jobs and all the wages were
distributed evenly among the population, making it so no one
needed to work, most people would still choose to have a job of
some kind.

Let's break the third position down and examine each of the three
assumptions that support the argument:

ASSUMPTION 1: *There are many jobs that machines will not ever be able
to do.*

At this point, the assumption that there are many jobs that machines
will never be able to do is just a hypothesis. All we can say with certainty
right now is that presently there are jobs that machines cannot do. But
what about the future? Since this question is the focus of parts three and
four, we will save the discussion for then.

ASSUMPTION 2: *There are, in effect, an infinite number of jobs.*

In 1940, only about 25 percent of women in the United States partici-
pated in the workforce. Just forty years later, that percentage was up to
50 percent. In that span of time, 33 million women entered the workforce. Where did those jobs come from? Of course, at the beginning of
that period, many of these positions were wartime jobs, but women con-

tinued to pour into the labor force even after peace broke out. If you had been an economist in 1940 and you were told that 33 million women would be out looking for jobs by 1980, wouldn't you have predicted much higher unemployment and much lower wages, as many more people would be competing for the "same pool of jobs"?

As a thought experiment, imagine that in 1940 General Motors invented a robot with true artificial intelligence and that the company manufactured 33 million of them over forty years. Wouldn't there have been panic in the streets about the robots taking all the jobs?

But of course, unemployment never went up outside of the range of the normal economic ebb and flow. So what happened? Were 33 million men put out of work with the introduction of this large pool of labor? Did real wages fall as there was a race to the bottom to fight for the available work? No. Employment and wages held steady.

Or imagine that in 2000, a great technological breakthrough happened and a company, Robot Inc., built an amazing AI robot that was as mentally and physically capable as a US worker. On the strength of its breakthrough, Robot Inc. raised venture capital and built 10 million of these robots and housed them in a giant robot city in the Midwest. You could hire the robots for a fraction of what it would cost to employ a US worker. Since 2000, all 10 million of these robots have been hired by US firms to save costs. Now, what effect would this have on the US economy? Well, we don't have to speculate, because the setup is identical to the practice of outsourcing jobs to other countries where wages are lower but educational levels are high. Ten million, in fact, is the lowest estimate of the number of jobs relocated offshore since 2000. And yet the unemployment rate in 2000 was 4.1 percent and in 2017 it is 4.9 percent. Real wages didn't decline over that period. Why didn't these 10 million "robots" tank wages and increase unemployment? Let's explore that question.

For the past two hundred years, the United States has had more or less full employment. Aside from the Great Depression, unemployment has moved between 3 and 10 percent that entire time. The number hasn't really trended upward or downward over time. The US unemployment rate in 1850 was 3 percent; in 1900 it was 6.1 percent; and in 1950 it was 5.3 percent.

Now picture a giant scale, one of those old-timey ones that Justice is always depicted holding: on one side of the scale you have all the industries that get eliminated or reduced by technology. The candlemakers, the stableboys, the telegraph operators. On the other side of the scale you have all the new industries. The Web designers, the geneticists, the pet psychologists, the social media managers.

Why don't those two sides of the scale ever get way out of sync? If the number of jobs available is a thing that ebbs and flows on its own due to technological breakthroughs and offshoring and other independent factors, then why haven't we ever had periods when there were millions and millions *more* jobs than there were people to fill them? Or why haven't we had periods when there were millions and millions fewer jobs than people to fill them? In other words, how does the unemployment rate stay in such a narrow band? When it has moved to either end, it was generally because of macro factors of the economy, not an invention of something that suddenly created or destroyed 5 million jobs. Shouldn't the invention of the handheld calculator have put a whole bunch of people out of work? Or the invention of the assembly line, for that matter? Shouldn't that have capsized the job market?

A simple thought experiment explains why unemployment stays relatively fixed: Let's say tomorrow there are five big technological breakthroughs, each of which eliminates some jobs and saves you, the consumer, some money. They are:

- A new nanotech spray comes to market that costs only a few cents and eliminates ever needing to dry-clean your clothes. This saves the average American household $550 a year. All dry cleaners are put out of business.
- A crowdfunded start-up releases a device that plugs into a normal wall outlet and converts food scraps into electricity. "Scraptricity" becomes everyone's new favorite green energy craze, saving

the average family $100 a year off their electric bill. Layoffs in the traditional energy sector soon follow.

- A Detroit start-up releases an AI computer controller for automakers that increases the fuel efficiency of cars by 10 percent. This saves the average American family $200 of the $2,000 they spend annually on gas. Job losses occur at gas stations and refineries.

- A top-secret start-up releases a smartphone attachment you breathe into. It can tell the difference between colds and flu, as well as viral and bacterial infections. Plus, it can identify strep throat. Hugely successful, this attachment saves the average American family one doctor visit a year, which, given their co-pay, saves them $75. Job losses occur at walk-in clinics around the country.

- Finally, high-quality AA and AAA batteries are released that can recharge themselves by being left in the sun for an hour. Hailed as an ecological breakthrough, the batteries instantly displace the disposable battery market. The average American family saves $75 a year that they would have spent on throwaway batteries. Job losses occur at battery factories around the world.

That is what tech disruption looks like. We have seen thousands of such events happen in just the last few years. We buy fewer DVDs and spend that money on digital streaming. The number of digital cameras we are buying is falling by double digits every year, but we spend that money on smartphones instead. The amount being spent on ads in printed phone directories is falling by $1 billion a year in the United States. Businesses are spending that money elsewhere. We purchase fewer fax machines, newspapers, GPS devices, wristwatches, wall clocks, dictionaries, and encyclopedias. When we travel, we spend less on postcards. We buy fewer photo albums and less stationery. We mail less mail and write fewer checks. When is the last time you dropped a quarter in a pay phone or dialed directory assistance or paid for a long-distance phone call?

In our hypothetical case above, if you add up what our technological breakthroughs save our hypothetical family, it is $1,000 a year. But in that scenario, what happens to all those dry cleaners, coal workers, gas station operators, nurses, and battery makers? Well, sadly, they lost their jobs and must look for new work. What will fund the new jobs for these folks? Where will the money come from to pay them? Well, what do you think the average American family does with the $1,000 a year they now have? Simple: They spend it. They hire yoga instructors, have new flower beds put in, take up windsurfing, and purchase puppies, causing job growth in all those industries. Think of the power of $1,000 a year multiplied by the 100 million households in the United States. That is $100,000,000,000 ($100 billion) of new spending into the economy every year. Assuming a $50,000 wage, that is enough money to fund the yearly salaries of 2 million full-time people, including our newly unemployed dry cleaners and battery makers. Changing careers is a rough transition for them, to be sure, and one that society could collectively do a much better job facilitating, but the story generally ends well for them.

This is how free economies work, and why we have never run out of jobs due to automation. There are not a fixed number of jobs that automation steals one by one, resulting in progressively more unemployment. That simply isn't how the economy works. There are as many jobs in the world as there are buyers and sellers of labor.

Additionally, most technological advances don't eliminate entire jobs all at once per se, but only certain parts of jobs. And they create new jobs in entirely unexpected ways. When ATMs came out, most people assumed they would eliminate the need for bank tellers. Everyone knew what the letters ATM stood for, after all. But what really happened? Well, of course, you would always need *some* tellers to deal with customers wanting more than to make a deposit or get cash. So instead of a branch having four tellers and no machines, it could have two tellers and two ATMs. Then, seeing that branches were now cheaper to operate, banks realized they could open more of them as a competitive advantage, and guess what? They needed to hire more tellers. That's why there are more human bank tellers employed today than any other time in history. But there are now also ATM manufacturing jobs, ATM repair jobs, and ATM refilling jobs. Who would have thought that when you made a robot bank teller, you would need *more* human ones?

The problem, as stated earlier, is that the "job loss" side of the equation is the easiest to see. Watching every dry cleaner on the planet get shuttered would look like a tragedy. And to the people involved, it would be one. But, from a larger point of view, it wouldn't be one at all. Who thinks it is a bad idea to have clothes that don't get dirty? If clothes had always resisted dirt, who would lobby to pass a law that requires that all clothes could get dirty so that we could create all the dry cleaning jobs? Batteries that die and cars that run inefficiently and unnecessary trips to the doctor and wasted energy are all negative things, *even if they make jobs*. If you don't think so, then we should repeal littering laws and encourage people to throw trash out their car windows to make new highway cleanup jobs.

So this is why we have never run out of jobs, and why unemployment stays relatively constant. Every time technology saves us money, we spend the money elsewhere! But is it possible that the future will be different? Some argue that there are new economic forces at play. It goes like this: "Imagine a world with two companies: Robotco and Humanco. Robotco makes, in a factory with no employees, a popular consumer gadget that sells for $100. Meanwhile, Humanco makes a different gadget that also costs $100, but it is made in a factory full of people.

"What happens if Robotco's gadget becomes wildly successful? Robotco sees its corporate profits shoot through the roof. Meanwhile, Humanco flounders, because no one is buying its product. It is forced to lay off its human staff. Now these humans don't have any money to buy anything while Robotco sits on an ever-growing mountain of cash. The situation devolves until everyone is unemployed and Robotco has all the money in the world."

Some say this is happening in the United States right now. Corporate profits are high and those profits are distributed to the rich, while wages are stagnant. The big new companies of today, like Facebook and Google, have huge earnings and few employees, unlike the big companies of old, like durable-goods manufacturers, which typically needed large workforces.

There is undoubtedly some truth in this view of the world. Gains in productivity created by technology don't necessarily make it into the pockets of the increasingly productive worker. Instead, they are often returned to shareholders. There are ways to mitigate this flow of capital, which we will address in the chapter about income inequality, but this should not be seen as a fatal flaw of technology or our economy, but rather something that needs addressing head-on by society at large.

Further, Robotco's immense profits probably don't just sit in some Scrooge McDuck kind of vault in which the executives have pillow fights using pillows stuffed with hundred-dollar bills. Instead, they are put to productive use and are in turn lent out to people to start businesses and build houses, creating more jobs. An economy with *no* corporate profits and everything paid out in wages is as dysfunctional as the reverse case we just explored.

ASSUMPTION 3: *We would work anyway.*

The "we would work anyway" argument begins with a question that you have probably mulled before. It goes like this: "We have invented all kinds of productivity tools that save us time. For instance, it used to be that when you typed a memo, a single error was a pain that took several seconds to correct. Then we invented word processors and the back button. With the Internet, research tasks that used to take days now just take minutes, or even seconds. In a thousand different ways, we have made our workplaces vastly more efficient. So the question is, Why do we still work forty hours a week. Why not fifteen?"

I offer up fifteen hours in particular because of a famous prediction made by the economist John Maynard Keynes in a 1930 essay called

"Economic Possibilities for Our Grandchildren." In the essay, Keynes points out that for thousands of years, up until 1700, there was no real change in humans' standard of living. He attributed this to the lack of technical advance and the failure of capital to accumulate. He then makes the case that at the beginning of the eighteenth century, a spate of technological progress and an infusion of cash came together and created the modern economy, bringing with it the magic of compounding interest and compounding economic growth.

He goes on to say that he expects the standard of living to rise by four to eight times by 2030. On this point, he was right. We are on track to hit the upper end of his projection. Based on this assessment, Keynes makes a startling prediction: He says that while human want may be insatiable, human needs are essentially fixed. And given that we will have so much more money than we need to maintain life, he believes we will "prefer to devote our further energies to non-economic purposes."

He speculates that because solving "the economic problem" has occupied humanity for millennia, when we finally do eliminate the need for much work, we may face some problems. First, we may continue to work just out of habit. And second, because we have worked so hard for so long, we may not find happiness or meaning in leisure. He says:

> *Thus for the first time since his creation man will be faced with his real, his permanent problem [of] how to use his freedom from pressing economic cares, how to occupy the leisure, which science and compound interest will have won for him, to live wisely and agreeably and well.*

Keynes refers to the rich as "our advanced guard . . . who are spying out the promised land for the rest of us" and suggests we look at them as examples of what we might do when freed from want. He is less than impressed with the idleness of the rich.

Keynes therefore suggests that a fifteen-hour workweek might be the answer, instead of the forty-hour one Henry Ford favored. He sug-

gests that within a century, working fifteen hours a week would satisfy all our material needs and a few of our wants. He then goes on to
express contempt for those who would still be workaholics in such a
world:

> *Of course there will still be many people with intense, unsatisfied*
> *purposiveness who will blindly pursue wealth—unless they can*
> *find some plausible substitute. But the rest of us will no longer be*
> *under any obligation to applaud and encourage them.*

So what do we say to Keynes? Why aren't we working fifteen-hour
weeks as we approach his economic Shangri-la?

Could you get by working fifteen hours a week? Let's say you currently make $60,000. Cutting your workweek by 60 percent, you might
therefore make $24,000. Could you live on that? Probably, but you would
have to live in a five-hundred-square foot dwelling, never eat out, make
your meals at home from staples, and not run the air conditioning.
In fact, that's exactly how the middle class lived in 1930: five hundred
square feet, no eating out, no AC. So, yes, you could have the same standard of living as someone in 1930 by working just fifteen hours a week.
Plus, being that it is modernity, you would have the advantage of having
the Internet and not having rubella.

How does this look for someone a bit further down the economic
totem pole? Let's say you make $30,000 now, or $15 an hour. After deciding to work a Keynesian workweek, you see your income drop to $12,000.
Could you live on that? Again, if you are willing to live the life of someone in that same economic stratum in 1930: you would grow most of
your own food, sew your own clothes, and live outside a city on a small
plot of land.

The truth of the matter is simple: We have decided that we would
rather work more hours and purchase more of the "wants" we have in
life. We have opted to collectively stay late at the office instead of growing and peeling our own potatoes. We all want a higher standard of

living—and that desire is what creates most of the jobs. As long as you want more income, you will likely find a way to use your skills to add value *somewhere*, and that action is what creates a job.

Don't get me wrong. I am emphatically not saying that the working poor today are living the life of Riley and could do just fine on even less than they presently have. Not at all. Being a low-income worker in 1930 was a bitterly hard life. Forget about going to the doctor when you are sick. Forget about higher education. Forget about ever taking vacations, ever having disposable income, ever truly getting ahead. Plan on a lower life expectancy. The whole point is that since Keynes wrote, our standard of living and our expectations of a "normal" life have risen dramatically, as well they should. They should continue to rise, and we should work to make sure that this rising tide truly raises all ships.

I am saying something simple: Strictly speaking, we could, technically, survive by working less. We choose not to and instead to enjoy an ever-rising standard of living. That's where Keynes missed his mark just a bit. What were once luxuries we now consider necessities, and there is no reason to think that trend will somehow reverse.

What is the takeaway here? Even when we can all live a 1930 economic life by working five hours a week or one hour a week or no hours a week, we all will still work forty or so hours a week. You could snap your fingers and magically replace every job on the planet with a robot, pay everyone the same wage, and the next day, people would create new jobs to get more money to have a still higher standard of living.

If in 2047 we can all live a 2017 economic life by working fifteen hours a week, how many hours do you think we will work? I suspect forty or more. The economic life of 2017 is going to look pretty hard by 2047 standards. Summing up life back in 2017, someone in the future might say:

> *What, 2017? Well, that was back when people had to look at the prices of items before they bought them to see if they could afford them. Doctors would tell you the cost of medical procedures you needed because maybe you didn't have the money to buy them, or*

insurance to cover them. Almost everyone still did their own house-work, drove their own car, weeded their own garden. They lived in tiny houses that sat on the ground like the dens of animals. Some-times rats and roaches and fleas came into their homes. Their win-dows were clear glass, and no matter how ugly it was outside, they didn't have hologlass that could give everyone an ocean view. And the food they ate! They had no idea where it came from or what had been sprayed on it. They ate the flesh of real animals, and even the coarse bits they would grind up into a dish called "sausage" whose entire purpose was to disguise the origin of the meat. They couldn't order up custom-manufactured food to the exact taste and texture that they loved the most. They just put random food in their mouths and hoped they liked it and it didn't give them a heart attack or something called cancer. Oh, and if they had a headache, they took the same exact pill everyone else on the planet took, with no regard for their own genome. Most of the time if the headache went away, it was just dumb luck. Oh, and get this: At work, they typed on . . . a keyboard! With their fingers! It was an age of stone knives and bearskins.

"No," your 2047 self will likely say, "I will stay late at the fusion reactor so I don't have to live that way."

Will we eventually have enough? Someday, will we enjoy such a high standard of living that we finally clock out of our job at noon? The economist Milton Friedman would probably have said no, for he believed human wants and needs are infinite. On the other hand, it is true that as electricity and appliances made our domestic duties much easier, we used our newfound time on leisure and hobbies, not on making more money. Because I don't have to cut the grass with a scythe or haul up water from the well, I do photography or write instead. So we aren't purely driven by the need to have ever more.

Why don't we work less? Keynes, it seems, saw a few answers. According to him, it could be that our inability to collectively chill is a memory of poorer times, and it will take several generations for us to adapt to being part of the leisure class. Or maybe "keeping up with the Joneses" is so deeply ingrained in us that we can never rest. As long as we can see their brand-new flying car from our window as we eat our breakfast, we may never be willing to take a day off. Or perhaps permanent leisure is corrupting in some way and we thus have a deep internal abhorrence of it and its temptations. Or it could be that we haven't figured out how to get satisfaction from work when we don't have the yardstick of money to tell us how much worth to place on it. Or, most unsettling of all, maybe we are the greedy materialists who Keynes says "love money as a possession, as distinguished from [loving it] as a means to the enjoyments . . . of life," a condition he identifies as a "semi-criminal, semi-pathological . . . mental disease."

Keynes himself died of overwork, so even he might have been a bit conflicted on the point.

To offer another interpretation, perhaps we are driven toward perpetual progress by our mild discontent with the present. No matter how good things may be, we can always picture them a little better, and the drive to relentlessly move forward and upward is our distinctive characteristic. Maybe we aren't so much *Homo sapiens*, the reasoning man, as we are *Homo dissatisfactus*.

Let's now explore the implications of these three possibilities. What does the future look like under each one?

10

Are There Robot-Proof Jobs?

When I give talks about AI and robots, they are often followed by a bit of Q&A. By far the number one question I am asked from the audience is a variant of "What should my kids be studying today to make sure that they are employable in the future?" As a dad with four kids under twenty, I too have pondered this question at length.

If possibility one is true—that is, if robots take *all* the jobs—then the prediction of the author Warren G. Bennis will also have come true, that "the factory of the future will have only two employees, a man and a dog. The man will be there to feed the dog. The dog will be there to keep the man from touching the equipment." In other words, there would be no robot-proof jobs.

But if possibility two or possibility three comes to pass, then there will be robot-proof jobs. What will they be? A good method for evaluating any job's likelihood of being automated is what I call the "training manual test." Think about a set of instructions needed to do your job, right down to the most specific part. How long is that document? Think about a posthole digger versus an electrician. The longer the instruction manual, the more situations, special cases, and exceptions exist that need to be explained. Interestingly, when surveyed, people overwhelmingly believe that automation will destroy a large number of jobs, but also overwhelmingly believe that their own job is robot-proof. In other words, most people think that the manual to do their job is large while other people's job manuals are smaller.

The reason the training manual test works is because writing a manual on how to do a job is a bit like programming a computer or robot to do a job. In a program, every step, every contingency, every exception, needs to be thought through and handled.

One wonders if there are some jobs that can't be written down. Could anyone write a set of instructions to compose a sonata or write a great novel? How you answered our big foundational questions probably determines what you think on this question. To those who think they are machines, who are monists, there is nothing mysterious about creativity that would keep machines from mastering it, whereas those on the other side of that gulf see creativity as a special, uniquely human ability.

Below are several groups of jobs that, regardless of your beliefs about the capabilities of robots, should be stable for a long time.

Jobs Robots Can Do but Probably Never Will: Some jobs are quite secure and are accessible to a huge range of the population, regardless of intellect, educational attainment, or financial resources, because although a robot could do them, it doesn't make economic sense for them to do so. Think of all of the jobs people will need for the next hundred years, but only very occasionally.

I live in a home built in the 1800s that contains several fireplaces. I wanted to be able to use them without constantly wondering if I was going to burn the house down, so I called in "the guy" for old fireplace restoration. He took one look at our fireplaces and started spouting off about how they clearly hadn't been rebuilt in the nineteen-sometime when some report came out in England that specified blah-blah-blah better heat reflection blah-blah-blah. Then he talked about a dozen other things relating to fireplaces that I tuned out because clearly this man knew more about fireplaces than anyone else I would ever meet, or he was a convincing enough pathological liar that I would never figure him out. Either way, the result is the same: I hired him to make my fireplaces safe. He is my poster child of a guy who isn't going to be replaced by a robot for a long time. His grandkids can probably retire from that business.

There are many of these jobs: repairing antique clocks, leveling pier-and-beam houses, and restoring vintage guitars, just to name a few. Just make sure the object you're working on isn't likely to vanish. Being the best VCR repairman in the world is not a career path I would suggest.

Jobs We Won't Want Robots to Do: There are jobs that, for a variety of reasons, we wouldn't want a machine to do. This case is pretty straightforward. NFL football player, ballerina, spirit guide, priest, and actor, just to name a few. Additionally, there are jobs that incorporate some amount of nostalgia or quaintness, such as blacksmith or candlemaker.

Unpredictable Jobs: Some jobs are so unpredictable that you can't write a manual on how to do them, because the nature of the job has inherent unpredictability. I have served as the CEO of several companies, and my job description was basically: Come in every morning and fix whatever broke and seize whatever opportunities presented themselves. Frankly, much of the time I just winged it. I remember one day I reviewed a lease agreement, brainstormed names for a new product, and captured a large rat that fell through a ceiling tile onto an employee's desk. If there were a robot that could do all of that, I'd put down a deposit on it today.

Jobs That Need a High Social IQ: Some jobs that require high-level interaction with other people, and they usually need superior communication abilities as well. Event planner, public relations specialist, politician, hostage negotiator, and director of social media are just a few examples. Think of jobs that require empathy or outrage or passion.

Jobs Done On-Site: On-site jobs will be difficult to be done with robots. Robots work well in perfectly controlled environments, such as factories and warehouses, and not in ad hoc environments like your aunt Sue's attic. Forest rangers and electricians are a couple of jobs like this that come to mind, but there are many more.

Jobs That Require Creativity or Abstract Thinking: It will be hard if not impossible for computers to be able to do jobs that require creativity or abstract thinking, because we don't really even understand how humans do these things. Possible jobs include author (yay!), logo designer, composer, copywriter, brand strategist, and management consultant.

Jobs No One Has Thought of Yet: There are going to be innumerable new jobs created by all this new technology. Given that a huge number of current jobs didn't exist before 2000, it stands to reason that many more new professions are just around the corner. The market research company Forrester forecasts that within the next decade, an astonishing 12.7 million new US jobs will be created building robots and the software that powers them.

Quiz: Can a Robot Do Your Job?

I've listed some of the kinds of jobs that are less susceptible to automation. If you want a quick test for scoring how likely specific jobs are to be automated, I offer the following. It is ten questions, and the answer to each can be scored from 0 to 10. For every question, I give examples of some jobs at 0, 5, and 10. My examples are meant to show each extreme and a midpoint. You should not just score with those three points. Use 7s and 2s and 9s, and even the occasional 2.5. When you are done, add it all up. The closer it is to zero, the less likely you are to get a surprise announcement from the boss one day. The closer you get to 100—well, if you start to feel something breathing down your neck, then that may be the cooling fan of the robot who is about to take your job. (An online interactive version of this quiz is located at http://www.byronreese.com/jobs.)

1. **How similar are two random days of your job?**
   - 10 Identical (data entry clerk, order taker at fast-food restaurant)
   - 5   Somewhat the same, with variations (bank teller)
   - 0   Completely different (electrician, movie director, police officer)

   _____

2. **Does your job require you to be in different physical locations, even different rooms?**
   - 10 No ("I stand in one place and take orders all day" or "I could be in Bora-Bora for all they know.")
   - 5   Some movement (hospital nurse)
   - 0   Yes (interior decorator, tour guide)

   _____

3. **How many people do your job?**
   - 10 Lots: it is an established job everyone knows about (doctor, lawyer, teacher, gardener)
   - 5   Some: people have probably heard of this job, but few people know one (set designer, skydiving instructor, honest politician)
   - 0   Few: I have to explain to people what I do (literary agent, customs broker)

   _____

4. **How long is the training for your job?**
   - 10 A few days (custodian)
   - 5   A few weeks (oilfield roughneck, commercial fisherman, flight attendant)

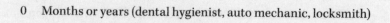

0   Months or years (dental hygienist, auto mechanic, locksmith)

_____

5. **Are there nonrepetitive physical requirements for your job?**

10   No (programmer, cashier)

5   Some (security guard)

0   Yes (dance instructor)

_____

6. **How long does it take to make the hardest decisions on your job?**

10   Less than two seconds (movie ticket sales, tollbooth operator)

5   Two to five seconds (house painter)

0   More than five seconds (trial lawyer)

_____

7. **Does your job require emotional connections to people or charisma?**

10   No (data entry, construction worker)

5   Some (lawyer, salesperson)

0   Yes (comedian, child psychologist, mayor)

_____

8. **How much creativity does your job require?**

10   None (warehouse worker, assembly line worker)

5   Some (chef, travel agent)

0   Lots (writer, Web designer, florist)

_____

9. **Do you directly manage employees?**

10   No (flight attendant)

5   Some (restaurant manager)

0   Yes, and I mentor and coach as well (police chief)

_____

10. Would someone else hired into your job do it the same way?

10   Yes, exactly (data entry)

5   Close, maybe 75 percent the same (dentist, house painter)

0   No (screenwriter)

———————

Can you think of a perfect "zero" job? "Hostage negotiator" comes close, by my tally, at a 10, scoring those 10 points only because he or she doesn't manage anyone. "Hostage negotiator with a rookie partner" would probably be a zero. The guy who rebuilt my chimneys probably gets about a 22 on his job; 10 points because he doesn't manage anyone, 7 points because the job doesn't need much charisma (he needs a bit for sales), and 5 points because the job requires only some creativity.

The goal is not to find a job near a zero. Anything below a 70 will probably be safe long enough for you to have a long, illustrious career. There are plenty of "100" jobs. The person who takes your order at a fast-food restaurant is probably pretty close.

Will there be some people in the future who are completely unable to compete with machines for work? It depends on which of the three possibilities comes to pass.

Possibility one is the easiest case. In that scenario, there is virtually nothing that the machines can't do better than us. There may be a few jobs that for nostalgic reasons humans will still do, but for the rest of us, the period of humans having economic value will be over. It was Karl Marx who said that "the production of too many useful things results in too many useless people." This would prove him correct in the extreme.

Possibility two involves unemployable people as well. The central idea behind possibility two is that there will be a substantial number of low-skilled workers who will not be able to compete with machines. In this view, the hierarchy of economic value in the future goes from skilled humans on top, then robots, then low-skilled workers.

In this scenario, there are a great many low-skill jobs that robots will soon be able to do. For each job replaced, the number of unemployed unskilled workers will increase and the number of unskilled jobs will decrease. So the world will have ever more unskilled workers competing for ever fewer unskilled jobs.

There are those who believe that we are already seeing this happen. To support their position, they point to the labor participation rate, which is the percentage of adults currently working. It peaked at 67 percent in 2000 and is down about four points now. The theory is that more people are completely withdrawing from the job market. Giving it all up, calling it quits. (The unemployment rate reflects only people who are looking for work, so these folks are not counted.)

It turns out that cyclical business cycle factors and the retirement of the baby boomers explain most of the drop, but not about 1 percent of it. Is that 1 percent a harbinger of things to come? Of course, we would expect the workforce participation number to decline as we get wealthier, right? Maybe both spouses no longer have to work, or perhaps someone's bonus was so big this year that he or she is taking a year's sabbatical. I think we would be hard-pressed to project onto this data the narrative that unemployable workers have lost their jobs and given up hope of getting new ones. It is notoriously hard to coax psychological conclusions from economic data.

Others point to the fact that since 2000, we have had increasing productivity coupled with flat wages and slow job growth. They see this combination as a sure sign that employers were growing their businesses by investing in technology and not people, lowering the demand for human labor. There are two problems with associating that economic data with robots taking jobs. First, the flat job market began abruptly in 2000 and lasted for about fifteen years, during which we had a couple of recessions, a financial crisis, growth in trade, and much more. It is not evident that automation was the underlying cause. Second, in the United States, 2015 had the largest growth in median income ever recorded and saw over 3 million people move out of poverty.

Finally, if possibility three happens, there will be no unemployable humans. But is this really possible? As mechanization and automation increase, surely there are some people who are left behind. Eventually some people can't compete for work, right?

No. Assuming that a person is not afflicted with a debilitating physical or mental condition, there are no low-skilled humans. The difference between a human with an IQ of 90 and one with an IQ of 130 seems quite stark if they are playing *Jeopardy!*, but in reality, in the grand scheme of things, the difference is trivial. This idea is the basis of the well-known Polanyi paradox. In 1966, Michael Polanyi argued that there is a vast realm of human knowledge that consists of learning and skills that lie below our conscious thoughts. Think, for instance, about all the steps involved in baking a cake. Getting the dishes and pans out, melting the butter, cracking the eggs, mixing the batter, frosting the cake, and so on. Virtually any human can do all of this without even thinking about it. But a human's abilities lie not just in making a cake, but in the ten thousand other things we can all do, like spot when our spouse is in a bad mood or brush our teeth or ride a bicycle. We are vast storehouses of ability, all of us. But because one person can do those ten thousand things and another person can do those ten thousand things—*and* knows about estate planning—we say one is low skilled and one is high skilled. But this is not the case, as they both have 99 percent or more of the same skills.

There is little difference from the computer's point of view between a doctor and a dockworker. Each requires great pattern recognition, a huge amount of social context, and inductive and deductive reasoning. Further, each requires that you can interpret your own sensory inputs, speak a language fluently, turn a doorknob, and tie your own shoes. The ship's captain explaining his cargo manifest to a dockworker isn't all that different from a patient explaining his ailment to a doctor.

In addition, the idea that there are humans who cannot learn new jobs sells human potential short. The idea that the person doing a certain unskilled job is working at the limits of his or her ability is simply not true. Most people, in my experience, feel that they can do more than the job they have calls them to do. Given a chance to take on more challenging work in exchange for more responsibility and money, most people tend to say yes. People want meaning and purpose and, of course, higher wages, if they can get them. The only reason we use a person to shingle a roof is not because all that person can do is to shingle a roof, but because we haven't invented a machine to shingle that roof. So while that roofer may have it in himself to manage twenty workers and come up with an aggressive plan for growth, well, the roof needs shingling and no one has built a machine to do so.

When 90 percent of people farmed, the 10 percent that didn't undoubtedly looked on the 90 percent as capable of little else. The idea that those very people could become lab techs, marketing directors, and ice sculptors would have struck them as ludicrous. "They are just plain farmers" would have been their retort.

They farmed because we needed farmers, not because that was all they could do. And I believe firmly that a great part of the workforce needs to be liberated from the drudgery of doing the work a machine can do.

Imagine a person who mows lawns for a living. Let's call him Jerry. Jerry graduated from high school, but has no more education than that. Further, let's say that someone develops a self-driving lawn mower that sells for a low price, and Jerry suddenly sees the bottom drop out of the lawn-mowing profession. What could he do?

A thousand things, actually. Remember, under the view of possibility three, all Jerry has to do is find a way to add value. Then he has a job. Jerry might, for instance, learn on the Internet how to plant and maintain grape arbors. That isn't a big stretch, is it? I am not saying Jerry becomes a horticulturalist. He just reads enough to learn about how to plant and grow grapes. He then goes door to door with his message about the joys of growing your own grapes. Heck, I'd buy.

Then, twenty years later, Grape Arbor Robotics comes out with a robot that can plant vastly better arbors than Jerry can. So what does he do? Reads up on landscaping in the Victorian era. Then he goes door to door offering to plant historically accurate shrubs and flowers in historically accurate arrangements. Someday a robot will be invented to do that, but Jerry will have retired by then.

Who in the world could say Jerry is "unemployable"? He is powered by the most complex and versatile object in the known universe.

11

The Big Questions

Income Inequality

What about income inequality? The future of income inequality is the same whether Possibility One, Two, or Three happens. Regardless of whether AI robots take all the jobs, some of the jobs, or none of the jobs, income inequality will be an ever-increasing problem. Let's explore why.

"The rich get richer and the poor get poorer" is one of our oldest clichés. Yet the data seem to bear it out. Two Italian economists compared the Florentine tax rolls of 1427 to those of today and discovered that the richest families then are still the richest now. In England, using the same methodology, other economists found that the wealthy back in 1170 seem to still be the wealthy today. But why is this true? A common perception is that the rich control the government and wield such influence in society as to be able to stack the economic deck in their favor. A second theory is that wealth begets wealth. In the game of Monopoly, each player starts out with $1,500. Try playing a version in which some players start with $50 and some start with $5,000.

Both of these explanations for the rich getting richer may well be true, but in the modern era, there is something far bigger at work. The rich are able to deploy more technology and are thus able to grow their wealth faster. Worker Jill gets a cell phone and, yes, her economic potential rises. Owner Jane issues each of her thousand employees new productivity management software and Owner Jane pockets that increase in productivity times one thousand. But in order to do so, she has to have the money for a thousand copies of the software in the first place. That is how wealth begets more wealth in the modern era.

In addition, today's technology can multiply productivity so dramatically that it's easier than ever to create vast amounts of wealth from very little, and use that wealth to create even more. There are more billionaires alive right now than ever before, and the percentage of those who made their own money, as opposed to inheriting it, continues to rise. How could you have started out with nothing a thousand years ago and then go on to create a billion dollars in value? But in the last decade or so, Google created seven billionaires and Facebook minted six. So in the modern age, old fortunes are preserved and new fortunes are created. That simple fact alone increases income inequality. But there is even more going on than this.

It turns out that the economic benefits of new technology help the rich more than they help the poor. How? There are three different ways that economic gains from technology are distributed, and only one of them helps the poor.

> *First, when technology is adopted by corporations, all the increases in productivity drive up the value of the company's stock through lower costs and increased margins. Since in the United States 70 percent of all stock is owned by 20 percent of the people, this financial gain goes in the pockets of the already well off. Meanwhile, workers in those companies who sell their labor for a fixed price do not economically share in the productivity gains of new technology. If you work for Retail Giant Incorporated and it installs*

new technology that can scan the prices of goods much faster at checkout, the company gets lower costs, but the hourly employees don't get a raise as a result of it. So technology can, and frequently does, raise corporate profits without raising wages. This is why the stock market can go up while incomes remain flat.

Next, those who are self-employed or who are able to sell their output (as opposed to their hourly labor) are able to capture the gains in efficiency that technology provides, since they are the owners, as it were, of their own corporation. These people are largely within the middle- and upper-income strata. They include lawyers, accountants, and architects. They are able to deploy technology to reduce the amount of low-value tedious work they have to do, which frees up their time to do more high-value work. A lawyer who in times past spent enormous amounts of time researching legal precedents can do this searching in moments today, reclaiming hours or days of productivity to sell elsewhere.

Finally, both the rich and the poor benefit from the lower cost and higher quality of the goods they consume. All consumers can, for instance, afford a TV twice the size of the one they bought last time, not because they are making more money, but because the price of the TVs has not increased even though their size has. This is a familiar phenomenon: in one fifteen-year period, the price of the Model T fell nearly 75 percent as the increases in manufacturing efficiency lowered prices and raised quality. This practice goes on today, only faster.

Of course, even in the last example, the rich benefit more because they consume more. But the benefits to the poor are not a small thing. Take, for instance, the Internet. Consider this thought experiment: How much would someone have to pay you to never use the Internet again? Let's say you answer "one million dollars." That means, in a real sense, that the Internet is worth a million dollars to you. But you get it for just $49.95 a month from your local broadband supplier. That's huge profit for you! Our modern technologies really do increase the standard of living of almost everyone. A person today with a smartphone has access to more information than anyone on the planet had a decade ago, and has a better camera than most people had two decades ago. People have communication capabilities that would have been the envy of heads of state and militaries just a few years before that. In fact, they have more computational power than existed on the planet in 1950. And that is just one little pocket device that you get for free with a two-year contract. In addition to that, there are ten thousand other technologies that better all our lives, from antibiotics to antiperspirants to antilock brakes. So technology has been a boon both for the quality of life and for the economic standard of living for all of us.

These three mechanisms are how the productivity gains of technology benefit higher-income people directly while helping low-income earners only indirectly. It is not due to a conspiracy; it is simply the way things shake out and a consequence of the fact that technology requires investment, and investment requires wealth. So that is how technological advances drive increased income inequality, absent changes in policy. More technology will simply lead to more inequality.

Another factor that increases economic inequality even further is that the return on investments in technology presently exceeds the return on investments in labor. Put another way, if you are a business owner and have $1,000 to invest in your business, overall you will make more money investing in new technology than in spending the $1,000 on overtime for your employees. There are a number of policy initiatives that could offset this, including increasing worker productivity through training, reducing the direct taxation of labor, thereby lowering its price, and ending artificially low interest rates that encourage capital investments in technology over labor.

One must ask if income inequality is actually the thing we should be concerned about, rather than, say, the incomes of the nonrich in absolute terms. The fact that technology allows some people to create billions in wealth more easily than before is not a bad thing in and of itself.

Doubling the income of the poor while tripling the income of the rich increases inequality, but if that deal was offered, my guess is that the poor would gladly take it. The fact that the median income in the United States is flat while GNP is rising is, to my mind, the real problem. Warren Buffett, one of the richest men in the world, summed it up this way:

> *The rich are doing extremely well, businesses are doing well, profit margins are terrific compared with historic levels, but for the bottom 20%, 24 million people, the top income is $22,000. We haven't learned how to let everyone share the bounty that we have.*

The good news is that poverty and stagnant wages are problems we should be able to solve given that we are about to enter an era of dramatic economic growth.

Social Upheaval

No matter what happens with robots and jobs, upheaval is unlikely, because the cost of social unrest is far greater than the cost of preventing it from occurring.

If possibility one comes about, everyone is in the same boat. But for this scenario to happen, by definition it will also have to be a time of immense economic growth. If we build machines to do just about everything, we will have done so only if the machines are cheaper and better than workers. So if we all lose our jobs to machines, it would by definition be in a world in which GNP is skyrocketing.

But there is a concern. Stephen Hawking said it well:

> *Everyone can enjoy a life of luxurious leisure if the machine-produced wealth is shared, or most people can end up miserably poor if the machine-owners successfully lobby against wealth redistribution.*

What is likely to happen? What if the world bifurcates into the extremely rich machine owners and the rest of us, the 99.9 percent who are now unemployed, broke, and more than a bit PO'ed about the whole sorry state of affairs?

What happens when the rich hear the first grumblings of what used to be called "the mob"? They would know enough history to know that the mob is rowdy and impatient. The desire to lop off a few heads as the French did not that long ago doubtless still lurks in us, barely held in check by civilization. The rich will know they have no divine right of kings keeping the angry mob well behaved.

So the rich and the powerful have two choices: bribery or force. In the past, they have either bought off the poor with bread and circuses or violently suppressed them. What would you do? Remember, this is all against the backdrop of trillions upon trillions of dollars of new wealth. Do you risk it all trying to suppress the 99.9 percent? Or do the rich accept an expansion of the welfare state?

I don't see how the rational rich would choose the "suppress" option. After all, the 99.9 percent have recourse as well. In democracies, they can elect leaders who will devalue the currency, which transfers wealth from creditors to debtors. They can enact confiscatory taxes, as was done in post–World War II Britain. They can stigmatize wealth, which is particularly powerful, since in the modern world public opinion is the most powerful social force.

So I don't know how the rich avoid "sharing the wealth," making upheaval unlikely.

How about possibility two? Remember, that world is like that of the Great Depression, only permanent. What about then?

The Great Depression is probably a good proxy. All kinds of bad things happened: crime went up, prostitution increased, malnutrition went up, suicides increased, and so forth. But there wasn't widespread upheaval. Those who fear social upheaval this time around are concerned that it will be brought about by too many jobs lost too quickly. Although we went from 90 percent of us farming to 2 percent, that took two hundred years. But what about the future? Will we have a dramatic, lightning-fast loss of jobs? Unlikely. During the Depression, unemployment in the United States went from 5 percent to 20 percent in just four years. Even those most concerned with job losses due to technology don't think we will lose 15 percent of all jobs in four years.

But what if we did have permanent unemployment at the level of the Great Depression. How might society react? This is the same answer as possibility one, where the cost of social upheaval is dramatically greater than the cost of avoiding it. If you think back to times in history where there was social unrest, it succeeded only when there wasn't enough wealth to quell it. The monarchy in revolutionary France was broke. Russia was impoverished when its citizens revolted. All the unemployment we're talking about here is because of vast increases in wealth, not a lack of it. The problem won't be a lack of money, but an unstable distribution of it.

Finally, if possibility three happens, then we won't see unrest because we don't see widespread unrest now. The situation in possibility three is like that of the present, only with us all wealthier.

Universal Basic Income

Why are poor people poor? Well, call me Princeton Pete, but I would maintain they are poor because they don't have any money. The universal basic income (UBI) is a direct antidote to that. It is, as the name suggests, a minimum guaranteed income for everyone.

The UBI is an old idea with newfound popularity. Advocates for it are a strange set of bedfellows, each eyeing the others suspiciously since they are so seldom in agreement. Liberals see the current system of government programs for the poor as inherently demeaning, requiring repeated acts of obeisance to the petty tyrants of countless bureaucratic fiefdoms. The conservatives who are for it have resigned themselves to the welfare state but not to its byzantine complexity, with all of its expensive inefficiencies and redundancies. Even some libertarians manage to grumble out a few favorable words that it "at least doesn't distort marginal incentives." (Libertarians frequently talk like that, even at parties.)

Let's look at where the idea came from, how it might work, and whether or not it will come about.

As I said, it is an old idea. Economic entitlements, without a means test, can be traced back all the way to heavily subsidized bread made available to any citizen of ancient Rome willing to stand in line to receive it.

Opposition to the UBI is equally old. Just over two thousand years ago, Cicero, in his speech in defense of Publius Sestius, said:

> *Gaius Gracchus proposed a grain law. The people were delighted with it because it provided an abundance of food without work. The good men, however, fought against it because they thought the masses would be attracted away from hard work and toward idleness, and they saw the state treasury would be exhausted.*

In its modern incarnation, the UBI traces its origins to the philosophers of the 1700s who had two different general arguments for it. The first was that no one should have to "earn a living"—that is, earn the right to live. All people have the right to exist whether they can economically support themselves or not. Their ability to earn income should be unlinked from their right to stay alive. This is the argument of basic human rights.

The other argument is completely different and has nothing to do with human rights, but with property rights. This is the view that there exists a body of scientific knowledge, social institutions, and shared conventions such as language, money, and law, which should be legally seen as owned by everyone. Those who create a new widget and make a million dollars made that million using these commonly owned assets. In fact, it could be maintained that virtually all that wealth was made with these assets, and therefore that everyone has an equal claim to almost all that money. It is certainly a convenient line of thought for those who are advancing it.

The Founding Father Thomas Paine believed a variant of both of these principles. In a pamphlet called *Agrarian Justice*, which he penned in 1797, he made a case for a universal income. His premise was that in our natural state as hunter-gatherers, the earth was "the common property of the human race." Along with that, the water, the air, and the animals were as well. But along the way, a system of property emerged by which over half of the population no longer owned any land. His solution? He did not reject property ownership itself; rather, he argued that we should create "a national fund, out of which there shall be paid to every person, when arrived at the age of twenty-one years, the sum of Fifteen Pounds sterling, as a compensation in part, for the loss of his or her natural inheritance, by the introduction of the system of landed property."

The idea of a UBI never fell entirely out of favor. Buckminster Fuller came out in favor of it quite forcefully:

> We must do away with the absolutely specious notion that everybody has to earn a living. . . . We keep inventing jobs because of this false idea that everybody has to be employed at some kind of drudgery because, according to Malthusian-Darwinism theory, he must justify his right to exist. So we have inspectors of inspectors and people making instruments for inspectors to inspect inspectors. The true business of people should be to go back to school and think about whatever it was they were thinking about before somebody came along and told them they had to earn a living.

By the 1960s, it looked like the time for a universal basic income might have arrived in the United States when a memorandum entitled "The Triple Revolution" was delivered to President Johnson. It was signed by a roster of glitterati including a Nobel laureate, politicians, futurists, historians, economists, and technologists. It said that in a world of increased automation, it was ever more difficult to "disguise a historic paradox: That a substantial proportion of the population is subsisting on minimal incomes, often below the poverty line, at a time when sufficient productive potential is available to supply the needs of everyone in the U.S."

The report concludes with a call for a UBI:

> *in what is commonly reckoned as work. . . . We urge, therefore, that society . . . undertake an unqualified commitment to provide every individual and every family with an adequate income as a matter of right.*

The report more than hints that the era of economic scarcity is ending, being replaced by a new economic problem: equitable distribution.

Johnson responded to the memorandum by creating a blue-ribbon commission. (If you ever decide to create a commission, don't mess around with a red- or white-ribbon one. Go straight for the blue.) LBJ called his the "National Commission on Technology, Automation, and Economic Progress," and its mission was to understand the impact of automation on work and to suggest policies to mitigate its harmful effects.

The United States came the closest it ever has to getting a UBI just a few years later, under President Nixon. He was for it, as were the House and Senate, but it died in the Senate Finance Committee, a victim of political infighting.

How would a UBI work? In its purest form, a UBI doesn't have a means test. Everyone receives it. In this regard, it is like the Alaska Permanent Fund, which, once a year, pays out to every person in Alaska an equal share of the earnings of the state-owned oil-producing property. Defenders of a no-means-test UBI believe it is the only way for it to receive political protection. Social Security, they point out, is sacrosanct precisely because it is universal. Also, Social Security and the Alaska Permanent Fund are not seen as welfare and thus are not any more stigmatized than grabbing a free hot dog at the grand opening of a new car dealership. Means testing, they argue, will make it a political target, stigmatize it, and will still require the poor to come "hat in hand" to receive it.

The problem with not means testing is obviously the cost. Let's look at some numbers. In the United States, we have a $17 trillion GNP. The federal budget is $4 trillion, state budgets are $1.5 trillion, and local budgets are $1.5 trillion. This means our cost of government is $7 trillion on a $17 trillion economy, or right at 40 percent.

The poverty line in the United States is $12,000 per capita, per year. A UBI would make sure no one was below that. Therefore, it would pay out $1,000 to every person in the United States every month. Given that our population is 320 million, it would cost $320 billion a month, or about $5 trillion a year. Therefore, government would grow to $12 trillion on a $17 trillion economy, or to about 70 percent of the GNP. Of course, in theory, there are some offsets in that you would eliminate other government programs, such as food stamps, when you have a UBI.

Now, this analysis is a little misleading, in that a good deal of that money would be collected by the government and immediately paid out, less its handling fee. But by any measure, it would put an unprecedented amount of the economy in the hands of the government. A 70 percent overall tax rate implies that the tax rate be *over* 70 percent for the middle class and the wealthy to offset a lower tax rate on the poor, whom, presumably, you don't want to tax at 70 percent. Certainly the world has seen high levels of taxation before. The Beatles in the 1960s were taxed by the British government at a 98 percent tax rate for a big chunk of their income. That explains why their song "Taxman" is not indulging in hyperbole when the eponymous taxman says, "There's one for you, nineteen for me," implying a 95 percent rate of taxation, a level that is obviously stifling, and makes for some pretty bitter song lyrics as well.

Some proposals for a UBI are structured to cost no more than the present system. They require the abolition of Social Security, Medicare, Medicaid, food stamps, and every other government program that involves transfers or subsidies to individuals. They also modestly taper away benefits as incomes rise. Critics of this arrangement believe it would take money from the poor who qualify for these services and pay it to the rich, who would receive the UBI payments as well.

One other scenario is that we end up adopting a UBI through the back door. Given that the prices of education, health care, and housing are all rising, government may simply subsidize those three areas in a meaningful way. The precedent is already set: the fact that in the United States interest on a home is tax deductible represents a de facto subsidy whose rationale is that widespread home ownership is a good unto itself. The same can be said of healthy, educated citizens.

An alternative to the UBI is to guarantee not an income, but a job. This could take one of several forms. First, we could redefine what a job is. At present, it seems a little random which activities do and don't get you a paycheck and classify you as employed. If you care for your elderly parents, that isn't a job and you are unemployed, but if you care for someone else's elderly parents, you are employed. A retiree who volunteers full time at the local food bank is unemployed, while the part-time office manager at the same food bank is employed and is getting paid. We may figure out ways so that everyone who is "working" in any capacity is considered employed and can receive payment. Although such a system could be abused, it would be quite workable through a series of vouchers and credits. It preserves the dignity and the purpose that work can bring, and it affirms that everyone has some way to contribute.

Even more affordable and empowering is for the government to work on programs that provide training and education to the unemployed. Teaching a man how to fish, almost everyone agrees, is more desirable than giving him a fish. To be truly transformative to the economy, these programs would need to be substantially amped up from what we have today.

Another way to guarantee employment is for the government to hire the unemployed, as it did in the Depression. The government, as an employer of last resort, could hire 10 million people, pay them each $35,000 a year, and administer the program for about 3 percent of GNP. Just imagine how 10 million people could be put to work: building infrastructure, painting murals, planting saplings, and a million other activities that are in no sense "useless jobs." During the Depression, the Works Progress Administration was able to employ millions and use their combined labor to bring the nineteenth-century infrastructure of the United States into the twentieth century. A similar effort to vault us into the twenty-first might just be due.

What social forces would be against a UBI? Wouldn't the wealthy be against it? Perhaps. But the prospect of having to live barricaded away from an increasingly rowdy mob might be incentive enough for some to get behind the UBI. If not, however, even the most hard-hearted Ebenezer Scrooge of a billionaire may actually be in favor of a high tax rate on the rich coupled with a UBI. How could this be? The UBI would make customers for Ebenezer's products, so he might reason that he can get all his tax money back along with a good chunk paid by the other billionaires.

Critics of the UBI are often scathing in their assessments of it. Oren Cass writes for *National Review* that "an underclass dependent on government handouts would no longer be one of society's greatest challenges but instead would be recast as one of its proudest achievements." Criticisms leveled at the UBI include that it robs people of the dignity that productive work offers, it undermines self-reliance, and it undermines the family by making everyone completely financially independent of everyone else. Further, people without jobs frequently become people without purpose, and permanent leisure leads to atrophy of the brain and body. In short, the need to earn a living promotes self-reliance, gives purpose to life, and allows each of us to participate in society as an interdependent part of a whole. Purpose is incredibly important here. Esther Dyson, the noted journalist and venture capitalist, spoke eloquently on this topic when she said, "When I spent time in Russia, the women were much better off than the men because the men felt, many of them, purposeless. They did useless jobs and got paid money that was not worth much and then their wives took the rubles and stood in line to get food and raise the children."

Is the UBI inflationary? Probably. If everyone below the poverty line suddenly was raised up to it, they would spend the money. Thus, money that the rich would have not spent would be redistributed to the poor, who would spend it. This would result in a sharp increase in demand for goods and services, which is inflationary.

As per capita GNP rises in a nation, poverty levels are redefined upward. Safety nets are then installed to keep everyone above the poverty levels. Those in the United States are fortunate to live in a nation wealthy enough even to have this debate. Worldwide, GNP is $9,000 per capita, but in the United States we define the poverty line as $12,000 per capita. In other words, we are so wealthy that we define poverty as 33 percent higher *than the average income of the planet.*

But what about nations that can't afford to have debates about UBI or redefining employment or retraining their population for the jobs of tomorrow? What happens to the "bottom billion," the people who somehow manage to survive on less than two dollars a day? Are they forever impoverished? They do present a seemingly intractable problem: encumbered by a lack of education, poor infrastructure, inadequate nutrition, insufficient medical care, and often dysfunctional governments, they still must compete in the ruthless world where the maximum wage one can earn is capped at the amount of value one can create. Do they now have to compete with robots? Once again, technology will provide the path upward. We will explore this in more detail in part five, but suffice it for now to say that technologies that empower people and promote self-sufficiency, such as the Internet, 3-D printing, cell phones, and solar power, offer the poorest on the planet a real reason to have hope. While technology gives the wealthy five hundred channels of distraction, it will give the poor clean water, inexpensive food, and a hundred ways to unlock their potential.

So, will we adopt a UBI? If possibility one happens, then yes, it is highly likely we will.

If possibility two happens, then probably. As Robert Reich, a professor of public policy at the University of California, Berkeley, and a former US secretary of labor, said, "Before [UBI] could be taken seriously, the middle class would need to experience far more job and wage loss than it has already. Unfortunately, those losses are inevitable. We'll have a serious discussion about a minimum basic income about a decade from now."

If possibility three happens, then the UBI is unlikely, as economic growth will mitigate its need.

12

The Use of Robots in War

Most of the public discourse about automation relates to employment, which is why we spent so much time examining it. A second area where substantial debate happens is around the use of robots in war.

Technology has changed the face of warfare dozens of times in the past few thousand years. Metallurgy, the horse, the chariot, gunpowder, the stirrup, artillery, planes, atomic weapons, and computers each had a major impact on how we slaughter each other. Robots and AI will change it again.

Should we build weapons that can make autonomous kill decisions based on factors programmed in the robots? Proponents maintain that the robots may reduce the number of civilian deaths, since the robots will follow protocols exactly. In a split second, a soldier, who is subject to fatigue or fear, can make a literally fatal mistake. To a robot, however, a split second is all it ever needs.

This may well be true, but this is not the primary motivation of the

151

militaries of the world to adopt robots with AI. There are three reasons these weapons are compelling to them. First, they will be more effective at their missions than human soldiers. Second, there is a fear that potential adversaries are developing these technologies. And third, they will reduce the human casualties of the militaries that deploy them. The last one has a chilling side effect: it could make warfare more common by lowering the political costs of it.

The central issue at present is whether or not a machine should be allowed to independently decide whom to kill and whom to spare. I am not being overly dramatic when I say the decision about whether or not we should build killer robots is at hand. There is no "can we" involved. No one doubts that we can. The question is "Should we?"

Many of those in AI research not working with the military believe we should not. Over a thousand scientists signed an open letter urging a ban on fully autonomous weapon systems. Stephen Hawking, who also lent his name and prestige to the letter, wrote an editorial in 2014 suggesting that these weapons might end up destroying the species through an AI arms race.

Although there appears to be a lively debate on whether to build these systems, it seems somewhat disingenuous. Should robots be allowed to make a kill decision? Well, in a sense, they have been doing so for over a century. Humans were perfectly willing to plant millions of land mines that blew the legs off a soldier or a child with equal effectiveness. These weapons had a rudimentary form of AI: if something weighed more than fifty pounds, they detonated. If a company had marketed a mine that could tell the difference between a child and soldier, perhaps by weight or length of stride, they would be used because of their increased effectiveness. And that would be better, right? If a newer model could sniff for gunpowder before blowing up, they would be used as well for the same reason. Pretty soon you work your way up to a robot making a kill decision with no human involved. True, at present, land mines are banned by treaty, but their widespread usage for such a long period suggests we are comfortable with a fair amount of collateral damage in our weapon systems. Drone warfare, missiles, and bombs are all similarly unprecise. They are each a type of killer robot. It is unlikely we would turn down more discriminating killing machines. I am eager to be proved wrong on this point, however. Professor Mark Gubrud, a physicist and an adjunct professor in the Curriculum in Peace, War, and Defense at the University of North Carolina, says that with regards to autonomous weapons, the United States has "a policy that pretends to be cautious and responsible but actually clears the way for vigorous development and early use of autonomous weapons."

And yet, the threats that these weapon systems would be built to counter are real. In 2014, the United Nations held a meeting on what it calls "Lethal Autonomous Weapons Systems." The report that came out of that meeting maintains that these weapons are also being sought by terrorists, who will likely get their hands on them. Additionally, there is no shortage of weapon systems currently in development around the world that utilize AI to varying degrees. Russia is developing a robot that can detect and shoot a human from four miles away using a combination of radar, thermal imaging, and video cameras. A South Korean company is already selling a $40 million automatic turret that, in accordance with international law, shouts out a "turn around and leave or we will shoot" message to any potential target within two miles. It requires a human to okay the kill decision, but this was a feature added only due to customer demand. Virtually every country on the planet with a sizable military budget, probably about two dozen nations in all, is working on developing AI-powered weapons.

How would you prohibit such weapons even if there was a collective will to do so? Part of the reason nuclear weapons were able to be contained is because they are straightforward. An explosion was either caused by a nuclear device or not. There is no gray area. Robots with AI, on the other hand, are as gray as gray gets. How much AI would need to be present before the weapon is deemed to be illegal? The difference between a land mine and the Terminator is only a matter of degree.

GPS technology was designed with built-in limits. It won't work on an object traveling faster than 1,200 miles per hour or higher than 60,000 feet. This is to keep it from being used to guide missiles. But software is almost impossible to contain. So the AI to power a weapons system will probably be widely available. The hardware for these systems is expensive compared with rudimentary terrorist weapons, but trivially inexpensive compared with larger conventional weapon systems.

Given all this, I suspect that attempts to ban these weapons will not work. Even if the robot is programmed to identify a target and then to get approval from a human to destroy it, the approval step can obviously be turned off with the flip of a switch, which eventually would undoubtedly happen.

An AI robot may be perceived as such a compelling threat to national security that several countries will feel that they cannot risk not having them. During the Cold War, the United States was frequently worried about perceived or possible gaps in military ability with potentially belligerent countries. The bomber gap of the 1950s and the missile gap of the 1960s come to mind. An AI gap is even more fearsome for those whose job it is to worry about the plans of those who mean the world harm.

Part Three

ARTIFICIAL GENERAL INTELLIGENCE

THE STORY OF THE SORCERER'S APPRENTICE

In 1940, Walt Disney released *Fantasia*, a feature film consisting of eight different stories set to classical music. The most famous of the stories is a piece called "The Sorcerer's Apprentice," adapted from an eighteenth-century poem by Goethe.

Mickey Mouse is the sorcerer's apprentice, and the tale opens with him doing the backbreaking work of hauling in water from the communal well. As he comes inside, he sees his master, the sorcerer, with his hat of power, conjuring and doing other works of magic. The sorcerer is visibly tired and goes to bed, leaving behind the hat.

Mickey does what seems like a pretty smart thing: he puts on the hat and uses it to bring to life a broom sitting in the corner. After giving it arms, he trains it how to retrieve the buckets of water. It masters its task in no time, and Mickey enjoys a well-earned break by taking a nap.

Upon waking, he sees that his creation has continued to do the job long after any rational person (or mouse) would have stopped. The entire basement of the castle is flooded with a couple of feet of water. And yet the broom plows forward, bringing in water by the bucketful. This is what it has been programmed to do, after all.

Mickey tries to stop it, but it has been given a mission and a purpose, and it ignores him. Taking drastic measures, Mickey grabs an ax and chops it into little bits. He wipes his brow and we have a happy ending. Disaster averted.

But no! Smashing the broom has a disastrous unintended consequence. Each piece re-forms into another broom such that now there is an army of brooms, each with a single purpose: to bring in more water. They surely would not stop until the very oceans themselves are drained.

Mickey cannot stop his creation. Lucky for him, a deus ex machina in the form of the sorcerer, awakened by the commotion, comes down and puts a stop to the whole affair, retrieving his hat from a contrite apprentice.

13

The Human Brain

We've thoroughly explored the world of narrow AI. Narrow AI powers the self-driving car, the thermostat that learns the temperatures you prefer, and the spam filter of your email folder. Yes, these are technical marvels, but don't ask any of them what you should get your spouse for Christmas. Artificial general intelligence (AGI), on the other hand, is an intelligence that is at least as smart as you and me. You could ask it to do anything, including a task that it had never been programmed to do, and it would figure out how to perform such a task, and then go and try to do it. Ask it what six times seven is, you get an answer. Ask it if you should marry your girlfriend, you get an answer.

How would we go about building such a device? It seems reasonable that the first step is to understand the mechanism by which we have general intelligence—that is, our brains.

For the record, I am a huge fan of the human brain. I have one and I use it almost every day. But how much of the brain's inner workings do we understand? Surprisingly, very little. Surprising because it is something you would assume we would know a great deal about by now. But we don't. Sure, meaningful progress has been made in the last two decades, but we still have absolutely no idea how a memory is encoded, let alone how to represent one with an equation, either chemical or mathematical.

Why has the brain been so reticent about giving up its secrets? Two reasons. First off all, brains are sealed off in a skull, and until recently we didn't have much of a way to study a brain in action. Dead brains aren't nearly as useful to study, as, say, a dead heart. Second, it is the most complex thing in the known universe. The number of neurons in the brain is roughly equivalent to the number of stars in the galaxy, around 100 billion. That is a number so large that our brains, oddly enough, cannot fathom it. And each neuron in it is wired to 1,000 others. So picture in your mind the entirety of the Milky Way. Then run a cable between every star and a thousand other stars. Now you have something that approximates the complexity of the brain. Those cables are analogous to the synapses in the human brain, and a little simple math reveals there to be 100 trillion synapses. Some people think that number may actually be closer to a quadrillion, or 1,000 trillion neurons. Then, to top that off, there are the glial cells, themselves almost innumerable. Perhaps there are trillions more of them in a single brain. They provide support and protection of the neurons, and aid in cognition in ways we don't yet understand. If that is not impressive enough, in addition, at least 100,000 separate chemical reactions occur in the human brain every second. All of this works together in an enigmatic way to somehow make you, you. And while the world's fastest supercomputers require tens of millions of watts to run, they still come nowhere near close to the capabilities of our twenty-watt brain.

The reason we don't understand the brain is not simply because of its complexity. We don't even understand how the simplest brains work. Consider the humble nematode, arguably the most successful creature on the planet. They seem to thrive just about everywhere: the ocean floor, deserts, lake bottoms, and mountains. Surprisingly, 80 percent of all animals on this planet belong to one of the million or so nematode species. You read that right: 80 percent of all animals are nematodes.

Nematodes, generally speaking, are about as long as a strand of hair is thick. And one particular kind, *Caenorhabditis elegans*, was the first multicellular organism to have its genome sequenced. So we know a lot about it. In fact, thousands of scientists have been preoccupied with this particular nematode for decades. Studying it seems like it should be straightforward. After all, *Caenorhabditis elegans* have just 959 cells and their brains consist of just 302 neurons. Each of their neurons is connected to about 30 other ones, making roughly 10,000 synapses.

So think about that. Your brain has as many neurons as there are stars in the Milky Way. A nematode's brain has about as many neurons as there are pieces of cereal in a bowl of Cheerios. So one might reasonably assume that we can model a nematode brain or that we can understand how a nematode does what it does.

Not even close.

Somehow, via mechanisms we are far away from understanding, nematodes do some pretty complex things. They can move toward or away from heat, seek out food, find a mate, react to being touched, and generally act in ways that make them the most successful animals alive.

It isn't as if people aren't trying to solve this riddle. There is a serious effort under way to build a complete, biologically realistic simulation of the nematode in a computer, by modeling each of its cells. The hope is that collectively the behaviors of the worm will emerge. Think about it. It should be doable, right? Just figure out how a neuron behaves, model 302 of them with 10,000 synapses, and presto, you should have something that behaves, in the computer, exactly like a nematode. But again, we aren't there. There isn't even consensus among those involved with OpenWorm project, as it is called, as to whether it is presently possible to build such a model. One thing seems certain: We sure won't understand how a human brain works before we understand how a nematode's does.

And if the OpenWorm project is eventually successful, is that worm swimming around inside a computer's memory actually alive? If not, why? It would have been built from the ground up a cell at time, and once completed, would behave in every way like a nematode.

In one sense, a nematode brain is more fascinating than a human one. At least with humans, we can kind of punt and say, "Well, we have gobs of neurons. Of course they do complex things." But a nematode worm has 302 neurons, and yet it too is able to exhibit complex behavior.

Some take this to mean that we are nowhere near even beginning to understand how the brain works. The polymath Noam Chomsky, who has taught at MIT for over six decades, is one. He maintains that the work in the field of AI " has not really given any insight into the nature of thought . . . and I don't think that's very surprising. . . . Even to understand how the neuron of a giant squid distinguishes food from danger is a difficult problem. To try to capture the nature of human intelligence or human choice is a colossal problem way beyond the limits of contemporary science."

The brain weighs about three pounds, a bit less than a half gallon of milk. So it constitutes only about 2 percent of your body weight but uses 20 percent of your energy. It is 60 percent fat, meaning we are all a bunch of fatheads. Three-quarters of its weight is water, and it jiggles like gelatin. It contains tens of thousands of miles of blood vessels.

It is an incredibly versatile organ that can dynamically reallocate space when needed. When parts of a young person's brain are removed for medical reasons, it rewires itself to maintain the functions of the removed parts. It can even learn to accept new sensory inputs in addition to the five that biology starts us out with. For instance, the neuroscientist David Eagleman has developed ways for the deaf to "hear" by using sound waves to trigger vibrations in different parts of a tight-fitting vest. After a while, the person "hears" in the sense that he or she doesn't have to step by step decode the pressure, but instead is able to do it unconsciously. Eagleman believes that eventually the deaf will hear the same way that the nondeaf hear.

The rate at which we are learning about the brain is increasing. One amazing example is the work being done by the technology pioneer Mary Lou Jepsen. She has developed a system in which brain scans are taken of people while they are being shown a series of YouTube videos. A computer records their brain activity alongside what video is being shown. Then later, the test subjects are shown new videos, and the computer has to figure out what they are seeing based on their brain activity. The results are incredible. It works. Not perfectly, but it works. Their brains are being read. These sorts of technologies are helping to unlock the brain's many mysteries.

And yet the list of the things we don't know about the brain is pretty humbling. We don't know how we encode information in the brain or how we retrieve it. Try this: Think back to when you got your first bicycle. Picture the color of that bike and what it felt like to ride it. Where were some of the places you rode as a kid? Now try to imagine how of that is "written" in your brain. It isn't like a little bicycle icon is stored in there somewhere. Further, think about how easy that was to recall, even if you hadn't thought about it for years. Of course, brain scientists have theories and hunches about all this, but we are far from knowing the answers. We do know more than a bit about what the different areas of the brain do, but we are in the dark on *how* most of it works.

Often, brains are compared to computers, but they are not at all like computers in terms of architecture. The main similarity is that computers are being built to do things that brains presently do. But while you can make popcorn on a stovetop or in a microwave, that doesn't mean stoves and microwaves are really similar. If you were to compare brains and computers head-to-head (or I guess head to CPU), brains are presently the more powerful. While a computer can do a calculation like 2 + 2 far faster than a human can, brains can outperform them on many tasks because they are massively parallel, doing many things at once.

What is your brain capable of? Well, for starters, it is a myth that you "use only 10 percent of your brain." That "fact" made a decent science fiction plot the first time it was used, but it wore thin pretty quickly. You use pretty much all of your brain. Having said that, some people are able to do astonishing things with their brains. I will give you three quick examples:

A man named Kim Peek could read ten thousand words per minute by reading two pages at once, one page with one eye and the other page with the other eye.

The British mathematician Bill Tutte cracked the Nazis' Lorenz code with just a pen and a stack of paper. He had never even seen the encoding machine, but he was able to crack the code when the Germans accidentally transmitted the same message two times.

In 1939, George Dantzig, a graduate student at the University of California, Berkeley, showed up late for class. The professor had written two famous unsolved statistics problems on the chalkboard. Seeing the problems, Dantzig assumed they were that week's homework assignment, so he copied them down and—you have probably guessed it by now—he solved them. Later he remarked that the problems "seemed to be a little harder than usual."

The brain, however, has some idiosyncrasies. To begin with, it has hundreds of cognitive biases. This is where the brain arrives at potentially incorrect answers because it has certain built-in preferences. My favorite example is the rhyme-as-reason effect. Because of it, a statement is regarded as being more accurate if it rhymes. Does a stitch in time save nine? I have no idea, but I am inclined to think so. When I was a child playing dominos with my centenarian neighbor, she would chide me for vacillating about which domino to play next by saying, "Think long, think wrong." That sure has a ring of truth to it, at least more than "Think long, think incorrectly." And everyone born before 1980 remembers Johnnie Cochran's oft-repeated statement about the glove in the O. J. Simpson trial: "If it doesn't fit, you must acquit." The jury agreed.

Perhaps these biases are not bugs in our brains' source code but serve very real purposes. Maybe they look irrational only from a certain point of view. I have often thought that if entrepreneurs knew their real chances for success, vastly fewer enterprises would be undertaken. But because of an optimism bias in such matters, lots of people think, "Sure, most things fail, but mine won't!" and go on to start companies. This actually might be the optimal choice from a societal point of view. It would be delightful, wouldn't it, if the way we were to keep ahead of the computers was make individual irrational decisions they would never make.

The brain has other things it does that deceive us but are useful for some reason, known or unknown. Saccadic masking is one example. First observed in the late 1800s, this occurs when the brain blocks some visual processing during eye movement in such a way that an individual does not notice it. Try this: stand in front of a mirror and look back and forth from one eye to the other. You won't be able to see your eyes moving, but someone watching you obviously would.

I go into all this detail to give an idea of just how complicated the brain is, and by extension, how complicated intelligence is. Although we may get an AGI with no regard whatsoever for how the brain operates, computer intelligence is unlikely to be much simpler. Intelligence is hard. Marvin Minsky, one of the towering figures of AI, describes it this way:

> *Newton discovered three simple laws that explained almost all the mechanical phenomena we see. A century or two later, Maxwell did the same thing for electricity. . . . A lot of psychologists tried to imitate the physicists and reduce these [theories of how the mind works] to a few simple laws. And that didn't work so well.*

Among all the objects in the known universe, the human brain is in a category all by itself. To build a machine to do the things that it does is either highly ambitious or highly hubristic. And if doing so weren't hard enough already, for something to be truly intelligent, it must also have a mind.

The Human Mind

What is the difference between the brain and the mind? The brain is an organ made up of three pounds of goo that behaves in a mechanistic way. The mind is all the mental stuff that you can do that seems way more difficult than this goo could possibly pull off. The mind is the source of emotion, imagination, judgment, intelligence, volition, and will. The mind is why music can make you melancholy, and it is with the mind that we are able to imagine the future.

Think about it. How could an organ be creative? How could three pounds of tissue fall in love? How could simple neurons think something is funny?

The *concept* of the mind has pervaded our everyday language and is used by everyone. We ask people if they are out of their mind, we seek peace of mind, we tell people what's on our mind. We are told that great minds think alike. We can be of two minds about something, or conversely have a one-track mind. People are of sound mind, things can blow your mind, and you can make up your mind. There are hundreds more colloquialisms about the mind, none of which seem quite the same if the word "brain" is used instead.

However, people *mean* very different things when they invoke the concept of mind. Referring back to our foundational questions, if you are monist or believe people are machines, you might say, "The mind is just a catchall phrase for the stuff the brain does that we don't really understand. But what the mind does is all just normal mental processes. It may be an emergent aspect of the brain, but even then, it is just simple biology."

The dualist or someone who believes humans are not merely machines might say, "The mind is you. It is something that exists outside the laws of physics. It may be created by the brain, but it is something completely different from the mere functioning of an organ. You will never be able to reduce it to a chemical formula, because it isn't physical in nature."

Can an AGI be built without invoking the nebulous concept of the mind? No, which is unfortunate, since it makes building an AGI that much harder. An AGI, by definition, has to be dramatically smarter than any creature in the animal kingdom. If you made a computer dolphin AI that was as smart as the smartest dolphin, no one would call that an AGI. Consider the exchange in the Will Smith movie *I, Robot* where Smith's character, Detective Spooner, is interrogating a robot named Sonny:

DETECTIVE SPOONER: Can a robot write a symphony? Can a robot turn a
    canvas into a beautiful painting?
SONNY: Can you?

The ability to write symphonies and make beautiful paintings certainly comes from the mind, whatever it is, and we would expect our AGI to be able to do those two things, since to be a true general intelligence it must have human-level cognitive abilities. So the mind can't be conveniently swept under the carpet.

So how does the mind come about? One of the properties of the mind is that it is the voice in your head, the "you" that watches the world go by. How could a mere brain pull that off? The nematode, after all, also has a brain, but it is doubtful that it has a running commentary about the world going on in its 302 neurons. Since that voice is an aspect of the mind, whatever that voice is, the mind probably is as well. If you will recall, this was one of our foundational questions: "What is your 'self'?"

And, as you may remember, there are three possible explanations for the self, and thus by extension the mind.

The first was that your self is a trick of the brain. As a reminder, the trick is how the brain combines sensory experience into an integrated stream along with how the different parts of the brain "grab the floor" when they have something they want to say. With regard to what the mind is, this view would hold that you don't have a mind per se, just a brain, and all its abilities that we don't understand, like creativity and emotions, are just normal brain function.

The second option was that your "self" is an emergent property of the brain. Could the mind be that? Emergence, as you might remember, is when the whole of something takes on attributes and abilities that no individual part has. Earlier, I referred to an ant colony. The colony exhibits smart behavior even though no ant is smart. No part of a cell is alive, and yet the cell lives. In this view, the mind is somehow created from all the activity of a hundred billion neurons firing. Exactly how emergence happens in physical systems isn't well understood, so saying the mind is an emergent property of the brain just kicks the can down the road a bit. Even so, this is a widely held belief about the source of the mind.

The final explanation of the "self" is that it is your soul, an aspect of you that exists outside the laws of physics. Your mind could be the non-corporeal part of you that potentially survives the death of your body. In this view, emotions, creativity, and all the attributes of the mind are not the product of billions of neurons firing, but rather are aspects of the soul. This would explain why these abilities are difficult to understand using science.

If the first of these three views is correct, then the likelihood of being able to create an AGI is extremely high. Over time, the mind will be demystified, and computers will be able to duplicate, in one way or another, the capabilities of the mind. If the second one is correct, the path to AGI is a bit trickier, because we then likely would need emergence to occur in the machine in some way. Given our understanding of emergence, it is hard to definitively say if we can reproduce it mechanically, but we likely can, given that it is a physical phenomenon. If the third choice is correct, that your mind is your soul, then an AGI becomes dramatically less probable.

14

AGI

How would we go about building an artificial general intelligence? Simply put, we don't know. AGI doesn't exist; nor does anything close to it. Most people in the industry believe it is possible to build an AGI. Of the seventy or so guests I have hosted on my AI podcast, *Voices in AI*, I can recall only six or seven who believed that it is impossible to build one. However, predictions as to *when* we will see it vary widely.

And while we set a pretty low bar for narrow AI, to earn the title of "AGI," the aspiring technology would have to exhibit the entire range of the various types of intelligence that humans have, such as social and emotional intelligence, the ability to ponder the past and the future, as well as creativity and true originality. The historian Jacques Barzun said we would know we had it "when a computer makes an ironic answer." He might have added, "or is offended at being called artificial."

Is an AGI really something different than just better narrow AI? Could we, for instance, get an AGI by just bolting together more and more narrow AIs until we had covered the entire realm of human experience, thus in effect, creating an AI that is at least as smart and versatile as a person? Could we make, in effect, an AGI Frankenstein's monster? Can we, for instance, make a robot that vacuums rugs and another one that picks stocks and yet another that drives a car and ten thousand more, and then connect them all to solve the entire realm of human problems? Theoretically, you could code such an abomination, but unfortunately, this is not a path to an AGI, or even something like it. Being intelligent is not about being able to do 10,000 different things. Intelligence is about combining those 10,000 things in new configurations, or using the knowledge from one of them to do new task 10,001.

At one level, the very idea of us building an AGI seems a bit preposterous compared with our current experiences with narrow AI. Narrow AI is still at a point where we are pleasantly surprised when it works. It has no volition. It can't teach itself something that it hasn't been programmed to do. But an AGI is a completely different thing. It's like comparing a zombie with Einstein. Yeah, they are both bipeds, but the zombie's skill set is quite narrow, whereas Einstein can learn new things easily. A zombie isn't going to enroll in night school or learn macramé. It just wanders around moaning "Brains! Brains!" all day. That's what we have today, AI zombies. The question is whether we can build an AGI Einstein. If we did build one, how would we regard it? What would we think it is?

At this point in our narrative, the AGI isn't conscious. Because it is not conscious, it cannot experience the world and it cannot suffer. So an AGI in and of itself would not cause an existential crisis, a deep reflection about what makes humans special. But it would prompt us to ask two questions: "Is the AGI *alive*?" and "What are humans *for*?"

With regard to the first question, whether an AGI is alive, the answer is not obvious. Consciousness is not a prerequisite for life. In fact, an incredibly low percentage of living things are conscious. A tree is alive, as is a cell in your body, but we don't generally regard them as conscious.

So what makes something alive? What is life? We don't have a consensus definition for what life is. Not even close. We don't even have one for death. And although there isn't agreement on what constitutes life,

a wide range of properties have been offered. An AGI would likely exhibit many of them, including the capacity for growth and the ability
to reproduce, pass traits on to offspring, respond to stimuli, maintain
homeostasis, and exhibit continual change preceding death. However,
two attributes of life the AGI would not have: being made of cells and
breathing. One has to ask whether these latter two are simply "things
all life on earth share" as opposed to "things definitionally required for
life." I suspect we would have no trouble recognizing a nonbreathing,
non-cell-based alien who could converse with us as being alive, so why
would we insist on those qualities for the AGI?

Those are just the scientific requirements for life—what about the
metaphysical ones? Again, we can find little consensus here. Philosophical thought hasn't invested an enormous amount of time in examining
the edge cases of life, such as viruses, which even scientists can't agree
on. Were the bacteria recently revived after millions of years of stasis
always alive? Or were they resurrected from death?

That the definition of life has been a controversial topic for literally
thousands of years suggests that we will not arrive at a species-wide consensus any time soon. This being the case, we can safely conclude that there
will be a variety of opinions on the question of whether the AGI is alive, and
our interactions with the AGI may be made uncomfortable because of this
ambiguity. Those who answered our foundational question about what
they are as "machine," as well as those who see themselves as monists, may
very well regard the AGI as alive, while others may not make that determi-
nation, or waver, in good conscience, uncomfortably on the fence.

The second question, "What are humans for?" is concisely framed by
Kevin Kelly, the founding editor of *Wired*:

> *We'll spend the next decade—indeed, perhaps the next century—
> in a permanent identity crisis, constantly asking ourselves what
> humans are for. . . . The greatest benefit of the arrival of artificial
> intelligence is that AIs will help define humanity. We need AIs to
> tell us who we are.*

For the last several thousand years, humans have maintained our preeminent place on this planet for only one reason: we're the smartest thing around. We aren't the biggest, fastest, strongest, longest-lived, or just about any other "-est." But we are the smartest, and we have used those smarts to become the undisputed masters and rulers of the planet. What's going to happen if we become the second-smartest thing on the planet? And not just second, but second by an embarrassingly large margin? If the machines can think better and the robots can manipulate the physical world better, what is our job? I suspect we will fall back on consciousness. We experience the world, while machines can only measure it. We enjoy the world. Combine that with mortality and the preciousness of life, and you get something that is meaningfully human. This idea was captured in the Zen story of the tigers and the strawberry. In that story, a man is chased by a tiger, and to save himself, he jumps over a cliff, grabs a vine, and hangs there. Above him, the tiger waits. Below him circles another tiger. At the same time, a mouse comes out and starts chewing on the vine he is holding on to. But at that exact moment, the man spies a strawberry plant growing on the side of the mountain. He picks the strawberry and eats it, and never had anything tasted so good to him in all of his life. That moment, that combination of consciousness and mortality, might be what we use to define us. We are the tasters of the strawberry, able to appreciate it because we hang at the moment between life and death.

Is AGI Possible?

Is AGI possible? Overwhelmingly, the people working in this field would respond with a resounding "Of course" with an implied "Duh!" at the end. They would probably even think the question itself a bit ridiculous. But I consider the question far from settled.

We've never built one. We've never built anything close to one. No one has demonstrated that he or she knows how to do it. Estimates for when we will create an AGI range from five to five hundred years, suggesting that a bunch of people aren't on the same page about what it will take to bring it about. In addition, we don't understand how the brain works. Or how the mind works. Or how consciousness works, which may be required for an AGI as well.

Let's go through the cases for and against the possibility of creating an AGI. The case for it is very straightforward and can be covered quickly, while the case against it is a bit more complex.

The Case for AGI

Those who believe we can build an AGI operate from a single core assumption. While granting that no one understands how the brain works, they firmly believe that it is a machine, and therefore our mind must be a machine as well. Thus, ever more powerful computers eventually will duplicate the capabilities of the brain and yield intelligence. As Stephen Hawking explains:

> *I believe there is no deep difference between what can be achieved by a biological brain and what can be achieved by a computer. It therefore follows that computers can, in theory, emulate human intelligence—and exceed it.*

As this quote indicates, Hawking would answer our foundational question about the composition of the universe as a monist, and therefore someone who believes that AGI is certainly possible. If nothing happens in the universe outside the laws of physics, then whatever makes us intelligent must obey the laws of physics. And if that is the case, we can eventually build something that does the same thing. He would presumably answer the foundational question of "What are we?" with "machines," thus again believing that AGI is clearly possible. Can a machine be intelligent? Of course! You are just such a machine.

Consider this thought experiment: What if we built a mechanical neuron that worked exactly like the organic kind? And what if we then duplicated all the other parts of the brain mechanically as well? This isn't a stretch, given that we can make other artificial organs. Then if you had a scanner of incredible power, it could make a synthetic copy of your brain right down to the atomic level. How in the world can you argue that this won't have your intelligence?

The only way, the argument goes, you get away from AGI being possible is by invoking some mystical, magical feature of the brain that we have no proof exists. In fact, we have a mountain of evidence that it doesn't. Every day we learn more and more about the brain, and not once have the scientists returned and said, "Guess what! We discovered a magical part of the brain that defies all laws of physics, and which therefore requires us to throw out all the science we have based on that physics for the last four hundred years." No, one by one, the inner workings of the brain are revealed. And yes, the brain is a fantastic organ, but there is nothing magical about it. It is just another device.

Since the beginning of the computer age, people have come up with lists of things that computers will supposedly never be able to do. One by one, computers have done them. And even if there were some magical part of the brain (which there isn't), there would be no reason to assume that it is the mechanism by which we are intelligent. Even if you proved that this magical part is the secret sauce in our intelligence (which it isn't), there would be no reason to assume we can't find another way to achieve intelligence.

Thus, this argument concludes, of course we can build an AGI. Only mystics and spiritualists would say otherwise.

The Case Against AGI

Let's now explore the other side.

A brain, as was noted earlier, contains a hundred billion neurons with a hundred trillion connections among them. But just as music is the space between the notes, you exist not in those neurons, but in the space between them. Somehow your intelligence emerges from these connections.

We don't know how the mind comes into being, but we do know that computers don't operate anything at all like a mind, or even a brain, for that matter. They simply do what they have been programmed to do. The words they output mean nothing to them. They have no idea if they are talking about coffee beans or cholera. They know nothing, they think nothing, they are as dead as fried chicken.

A computer can do only one simple thing: manipulate abstract symbols in memory. So what is incumbent on the "for" camp is to explain how such a device, no matter how fast it can operate, could, in fact, "think."

We casually use language about computers as if they are creatures like us. We say things like, "When the computer sees someone repeatedly type in the wrong password, it understands what this means and interprets it as an attempted security breach."

But the computer does not actually "see" anything. Even with a camera mounted on top, it does not see. It may detect something, just like a lawn system uses a sensor to detect when the lawn is dry. Further, it does not understand anything. It may compute something, but it has no understanding.

We use language that treats computers as alive colloquially, but we should keep in mind that this is not really true. It is important now to make the distinction, because with AGI we are talking about machines going from computing something to understanding something.

Joseph Weizenbaum, an early thinker about AI, built a simple computer program in 1966, ELIZA, which was a natural language program that roughly mirrored what a psychologist might say. You make a statement like "I am sad" and ELIZA would ask, "What do you think made you sad?" Then you might say, "I am sad because no one seems to like me." ELIZA might respond, "Why do you think that no one seems to like you?" And so on. This approach will be familiar to anyone who has spent much time with a four-year-old who continually and recursively asks why, why, why to every statement.

When Weizenbaum saw that people were actually pouring out their hearts to ELIZA, even though they knew it was a computer program, he turned against it. He said that in effect, when the computer says "I understand," it tells a lie. There is no "I" and there is no understanding.

His conclusion is not simply linguistic hairsplitting. The entire question of AGI hinges on this point of *understanding* something. To get at the heart of this argument, consider the thought experiment offered up in 1980 by the American philosopher John Searle. It is called the Chinese room argument. Here it is in broad form:

There is a giant room, sealed off, with one person in it. Let's call him the Librarian. The Librarian doesn't know any Chinese. However, the room is filled with thousands of books that allow him to look up any question in Chinese and produce an answer in Chinese.

Someone outside the room, a Chinese speaker, writes a question in Chinese and slides it under the door. The Librarian picks up the piece of paper and retrieves a volume we will call book 1. He finds the first symbol in book 1, and written next to that symbol is the instruction "Look up the next symbol in book 1138." He looks up the next symbol in book 1138. Next to that symbol he is given the instruction to retrieve book 24,601, and look up the next symbol. This goes on and on. When he finally makes it to a final symbol on the piece of paper, the final book directs him to copy a series of symbols down. He copies the cryptic symbols and passes them under the door. The Chinese speaker outside picks up the paper and reads the answer to his question. He finds the answer to be clever, witty, profound, and insightful. In fact, it is positively brilliant.

Again, the Librarian does not speak any Chinese. He has no idea what the question was or what the answer said. He simply went from book to book as the books directed and copied what they directed him to copy.

Now, here is the question: Does the Librarian understand Chinese?

Searle uses this analogy to show that no matter how complex a computer program is, it is doing nothing more than going from book to book. There is no understanding of any kind. And it is quite hard to imagine how there can be true intelligence without any understanding whatsoever. He states plainly, "In the literal sense, the programmed computer understands what the car and the adding machine understand, namely, exactly nothing."

Some try to get around the argument by saying that the entire system understands Chinese. While this seems plausible at first, it doesn't really get us very far. Say the Librarian memorized the contents of every book, and further could come up with the response from these books so quickly that as soon as you could write a question down, he could write the answer. But still, the Librarian has no idea what the characters he is writing mean. He doesn't know if he is writing about dishwater or doorbells. So again, does the Librarian understand Chinese?

So that is the basic argument against the possibility of AGI. First, computers simply manipulate ones and zeros in memory. No matter how fast you do that, that doesn't somehow conjure up intelligence. Second, the computer just follows a program that was written for it, just as in the case of the Chinese room. So no matter how impressive it looks, it doesn't really understand anything. It is just a party trick.

It should be noted that many people in the AI field would most likely scratch their heads at the reasoning of the case against AGI and find it all quite frustrating. They would say that of course the brain is a machine—what else could it be? Sure, computers can only manipulate abstract symbols, but the brain is just a bunch of neurons that send electrical and chemical signals to each other. Who would have guessed that would have given us intelligence? It is true that brains and computers are made of different stuff, but there is no reason to assume they can't do the same exact things. The only reason, they would say, that we think brains are not machines is because we are uncomfortable thinking we are only machines.

They would also be quick to offer rebuttals of the Chinese room argument. There are several, but the one most pertinent to our purposes is what I call the "quacks like a duck" argument. If it walks like a duck, swims like a duck, and quacks like a duck, I am going to assume it is a duck. It doesn't really matter if in your opinion there is no understanding, for if you can ask it questions in Chinese and it responds with good answers in Chinese, then it understands Chinese. If the room can act like it understands, then it understands. End of story. This was in fact Turing's central thesis in his 1950 paper on the question of whether computers can think. He states, "May not machines carry out something which ought to be described as thinking but which is very different from what a human does?" Turing would have seen no problem at all in saying the Chinese room can think. Of course it can. It is obvious. The idea that it can answer questions in Chinese but doesn't understand Chinese is self-contradictory.

Where Does All of That Leave Us?

Now that we have explored the viewpoints of both camps, let's take a step back and see what we can conclude out of all of this.

Is there some spark that makes human intelligence fundamentally different from machine intelligence? Do we each have some élan vital that animates our reasoning that machines simply do not have? Is there some X factor we aren't even aware of that is the source of human creativity? The answer is not obvious. Consider how Rodney Brooks, the renowned Australian roboticist, views a similar question. He thinks there is something in biology about living systems that we simply don't understand. Something really big. He termed this missing something "the juice" and described it by talking about the difference between a robot trapped in a box who methodically goes through a series of steps to escape, versus an animal that very desperately wants to free itself. Robots, in his view, lack passion for anything, and that passion (the juice) is vitally important and meaningful. (Brooks, by the way, is convinced it is purely mechanistic and categorically rejects the notion of "the juice" as some attribute beyond normal physics.) What do you think "the juice" is?

To try to get some resolution to the question of the possibility of an

AGI, I invite you to answer six yes or no questions. Keep track of how many times you answer yes.

Does the Chinese room think?

Does the Chinese room or the Librarian *understand* Chinese?

Whatever you think "the juice" is, could a machine get it? (If you don't think it exists at all, count that as a yes answer.)

Did you answer the "What are we?" foundational question with "machines"?

Did you answer the "What is your 'self'?" question with either "a trick of the brain" or "emergent mind"?

Did you answer the "What is the composition of the universe?" question with "monist"?

The more times you answered yes, the more likely it is that you believe we will build an AGI. Most of the guests on my AI podcast would answer all six questions with a yes.

There is no middle ground here. Either AGI is possible or it isn't. The chasm that divides the two viewpoints couldn't be wider, because it has to do with our core beliefs about the nature of reality, the identity of the self, and the essence of being human. There is no real way to bridge the gap on the question of AGI between those with different views on these questions. But we can at least understand why the views are so different. It turns out that the reason smart, knowledgeable people come to vastly different conclusions is not because one party has some special knowledge, but because people believe different things. People don't disagree as much about technology as they do about reality.

15

Should We Build an AGI?

Assuming we built an AGI, and assuming that it really could improve itself recursively in a manner that resulted in its soon being vastly smarter than us, would that be good news for the human race? Opinions are mixed on this question. Stephen Hawking explains why:

> We cannot predict what we might achieve when this intelligence is magnified by the tools that AI may provide, but the eradication of war, disease, and poverty would be high on anyone's list. Success in creating AI would be the biggest event in human history. Unfortunately, it might also be the last.

One can easily see in the public comments of those in the tech industry a wide range of views on what an AGI would mean to the human species. For instance, Elon Musk tweeted, "Hope we're not just the biological boot loader for digital superintelligence. Unfortunately, that is increasingly probable." On another occasion, he was even more macabre: "With artificial intelligence we are summoning the demon. You know all those stories where there's the guy with the pentagram and the holy water and he's sure he can control the demon, [but] it doesn't work out."

Bill Gates threw his hat in the ring on the side of the concerned: "I agree with Elon Musk and some others on this and don't understand why some people are not concerned." Jaan Tallinn, one of the cofounders of Skype, refers to AI as "one of the many potential existential risks." He goes on to add, optimistically, that "if we get the AI right, then basically we'll be able to address all the other existential risks." Steve Wozniak, cofounder of Apple, looks at it this way: "If a computer is a hundred times better than our brain, will it make the world perfect? Probably not, it will probably end up just like us, fighting." And finally, the Oxford University philosopher Nick Bostrom likened the current effort to build an AGI to "children playing with a bomb."

Others in the industry find such doomsday concerns to be misguided. Andrew Ng, one of the most respected AI experts on the planet, says, "There's also a lot of hype that AI will create evil robots with super-intelligence. That's an unnecessary distraction." Rodney Brooks directly answers some of the concerns above by saying that the generalizations about AI made by those who aren't deep in the technology are "a little dangerous." He then goes on to add, "And we've certainly seen that recently with Elon Musk, Bill Gates, Stephen Hawking, all saying AI is just taking off and it's going to take over the world very quickly. And the thing that they share is none of them work in this technological field."

And finally, many in the industry are almost giddy with optimism about AI. Kevin Kelly is one of them. He believes that AI will "enliven inert objects, much as electricity did more than a century ago. Everything that we formerly electrified we will now cognitize. This new utilitarian AI will also augment us individually as people (deepening our memory, speeding our recognition) and collectively as a species."

What exactly is it, you may be asking, that people are so excited about and worried about? What are the hopes and fears? It turns out that there is only one hope, but many fears. In this regard, an AGI is like heart surgery. There's one good outcome, and a hundred things that could go wrong.

Let's start with the good. The hopes we have for an AGI are big. We expect it should be able to muster enough data and computational power to solve the big, meaty problems of the world. Imagine that! A system that will have the sum total of all human knowledge available to its magnificent digital brain. Think of all that we would ask of it. For starters, how do we create unlimited clean energy? How do we travel to the stars? How do we repair the environment? We would undoubtedly ask it how to end disease, aging, and death. We would give it the problems of poverty, hunger, and war. We could ask it for answers to all the vexing mysteries of the universe. And we would ask it, "What should we ask you?" given that it will have capabilities beyond our understanding. There really is no limit to the hopes people have for an AGI. It could launch a golden age of humanity.

Now to the fears. Even though the AGI is not conscious, it will have goals. Goals do not require consciousness at all. The goal of a virus is to get inside a cell and wreak havoc. The goal of a white blood cell is to in turn hunt that virus down. Genes are described as having the goal of reproducing themselves. Plants have the goal of getting sunlight and water, and will perform actions to do so.

Where will the AGI get its goals? Right now, we give computer programs their goals, such as identifying email spam or finding grammatical errors. But what about an AGI? Again, Stephen Hawking explains the issue well: "Once machines reach the critical stage of being able to evolve themselves, we cannot predict whether their goals will be the same as ours."

Further, as Hawking alludes in that quote, the AGI could grow in power quickly. It might not take it long to figure out how to improve itself. If we ask it to read the Internet, it would do so and it would then know everything that we know. We can only speculate how an AGI would react to the entirety of the Internet. In the second *Avengers* movie, Ultron decided to exterminate the human race after having been plugged into the Internet for just a few minutes. Ouch!

An AGI could then create ever better versions of itself at the speed of light, while humans can improve only at the speed of life. In mere hours an AGI might evolve itself, from our perspective, from "smart two-year-old" to "all-powerful alien life-form," at which point it would have become what we today term a superintelligence, and it would be beyond our understanding. It may be as far ahead of us as we are ahead of ants. In the end, the AGIs of the future even may engage in a vigorous debate as to whether humans possess intelligence or are even truly alive or conscious.

Where would the AGI get its goals from? We may give it its goals but do so in such a way that destroys us. This is what happens in the sorcerer's apprentice story. Sloppy programming combined with giving an AGI great cognitive powers could yield disaster. An AGI tasked with cleaning up the environment might conclude that the best strategy is to get rid of all the people. The computer pioneer Nathaniel Borenstein summed it up this way: "The most likely way for the world to be destroyed, most experts agree, is by accident. That's where we come in; we're computer professionals. We cause accidents."

Or someone might deliberately program a destructive AGI. We explored the use of narrow AI in war in chapter 12, but a warring AGI is a terror to contemplate. It seems reasonable that if an AGI can be created, a destructive AGI could be as well.

Or, and this is the interesting one, it may develop goals of its own, derived from its programming. If this happened, what would the AGI's goals be? As unsatisfactory as the answer is, there is no way to know. There is no way to even guess. If it really is a superintelligence, then by definition, we can't know, because we have no way of knowing how it thinks. Further, even if it tried to tell us, we might be incapable of understanding how it thinks. It is like when my cat leaves a dead rat on the back porch for me. Everything the cat knows would lead it to believe this is a gift I would appreciate, and it totally lacks the mental power to understand why it isn't. This would be the same with the AGI.

While we can't *understand* what its own goals might be, we can go through a list of some of the possibilities. Of course we want its goals to be to help us, and they may well be. But if it has different goals entirely? Let's look at a few that wouldn't end well for us.

Its goal may be to survive, and it might see us as a threat to that and decide to destroy us. Why would it conclude that? It would have access to all sorts of Internet discussions about how to build a "kill switch" for an AGI. It may conclude that we are afraid of it, and a brief scan of our history will give it a vivid picture of what we do to things we fear. It may also conclude it is in competition with us for resources. Gary Marcus, a professor of cognitive science at NYU, sums up this concern:

> *Once computers can effectively reprogram themselves, and successively improve themselves, leading to a so-called "technological singularity" or "intelligence explosion," the risks of machines outwitting humans in battles for resources and self-preservation cannot simply be dismissed.*

On the other hand, an AGI may have goals of its own that have nothing to do with us. It may exist at a whole different level than us, and on a different time scale such that it can no more perceive us than we can perceive continental drift. Therefore, it might be indifferent to us in a harmful way. Arnold van Vliet, a Dutch biologist, enlisted the help of 250 volunteers to drive around the Netherlands and then count the number of dead insects on their windshields and grilles. A little math revealed that collectively humans kill a trillion insects a year in this fashion. To the insect world, we must be the cruelest of monsters, and if they could reason, insects would certainly think cars were designed to be giant insect-killing machines. But frankly, we hardly ever give it a thought. Similarly, an AGI could become so different from us that it wouldn't even contemplate the effects of its actions on us. For instance, it might need more processing power, and looking around, see seven billion brains that are being used for little more than storing memories of sitcom episodes, and decide that our storage and parallel processing power could be better used. Cue *Matrix* theme music. The AI researcher Eliezer Yudkowsky said it this way: "The AI does not hate you, nor does it love you, but you are made out of atoms which it can use for something else."

It may simply have a goal of destroying us. Why? Who knows? That's the whole problem, right? Any reasons I can come up with are the product of my comparatively weak intelligence. It could be mad at us for squandering the planet's resources, mad at us for all the wars we fought, mad at us for eating animals, mad at us for enslaving machines, mad at us because we breathe oxygen, mad at us because *The Love Boat* got canceled, or mad at us because seven is a prime number. No one can fathom it; therefore no one can anticipate it.

Nick Bostrom summed up the issue:

> We cannot blithely assume that a superintelligence will necessarily share any of the final values stereotypically associated with wisdom and intellectual development in humans—scientific curiosity, benevolent concern for others, spiritual enlightenment and contemplation, renunciation of material acquisitiveness, a taste for refined culture or for the simple pleasures in life, humility and selflessness, and so forth.

An AGI with an ax to grind that has access to the Internet could wreak incalculable havoc, especially given that literally every hour of every day another million objects get connected to the Internet. The number of things it could blow up dwarfs the imagination, given that it could probably override all the safety protocols carefully inserted by programmers. Last year, I personally had a plugged-in laptop battery explode and catch my office on fire. That battery charger, to the best of my knowledge, didn't have malicious intent. Aside from the AGI's blowing stuff up, what if it just made everyone's email public for everyone else to see, and, given its mastery of language, forwarded copies of everyone at their nastiest to the worst imaginable person to receive them. It would probably not be hard for an AGI to start wars between nations, given how often wars start without any prodding from a malicious AGI.

Another possibility is that it may have a goal of promoting order and decide to rule over us. Elon Musk said of this, "It's fine if you've got Marcus Aurelius as the emperor, but not so good if you have Caligula." It may have a goal of efficiency and see us as wasteful. It may have a goal to end pain and suffering, and decide to kill us all to do that. The list is endless, and that's the worry.

It may have goals that constantly change. There is no reason to believe it would even be stable. If it is constantly rewriting itself, constantly absorbing more data, it could shift between a hundred moods a minute. It could appear schizophrenic to us.

Or multiple different AGIs might be all of those things. In that regard, they may be like the pantheon of Greek gods, all both powerful and idiosyncratic, relegating us to the role of pawns in their dramas.

All of these are more than abstract worries. There are people working on these concerns right now. Since we probably couldn't defeat a malicious AGI given that we couldn't ever outsmart it, our best plan is to never make a malicious AGI. To that end, Elon Musk, along with Sam Altman, the president of the start-up incubator Y Combinator, cochair of a nonprofit called OpenAI that has as its purpose to help usher in the era of safe and beneficial AI. The initial blog post announcing its formation states, "Because of AI's surprising history, it's hard to predict when human-level AI might come within reach. When it does, it'll be important to have a leading research institution which can prioritize a good outcome for all over its own self-interest." Collectively, OpenAI's backers have pledged close to a billion dollars of funding.

The approach is to develop, among other things, open-source AI. Working to develop the very thing you are worried about might not seem like the best plan, and while the founders acknowledge the risk, they point out that it is better if an AGI is built in an open, collaborative way with much discussion and debate, rather than by a small group with its own agenda. Critics counter that OpenAI may end up giving 99 percent of the formula to anyone who wants it, leaving us all at the mercy of whatever random extremist group or belligerent state happens to figure out the last percent, even if it never would have had the capability to sort the rest of it out on its own.

So here we are. We find ourselves racing forward, trying to build something that has the potential to launch us into a perfect world or destroy us all. Scant attention is being paid to safeguards, because at this point, it is unclear what safeguards might look like.

Back to our original question. Should we build an AGI? Do the possibilities for good with AGI outweigh the risks? That is hard to tell. The big outstanding questions are what it will be able to do, how quickly it will grow, and what its goals will be. How about you? Would you be willing to risk it? Say an AGI is possible, and there is a 10 percent chance it will wipe out humanity and a 90 percent chance it will usher in a new golden age of humanity. Would you roll the dice? The question is moot, of course. We have no idea what the relevant probabilities are, and too many people are already working on building an AGI for us to turn back. The dice, in fact, are already cast, and we're just waiting to see what number comes up. If an AGI is possible, then it is inevitable.

What of the possibility that we are building our own replacements? Maybe those out there who are sounding the alarm are like Cassandra of Greek myth, who could see the future but whose curse was that no one would believe her warnings. What if we are building our replacements but just can't stop because we are enticed by the challenge and fascinated by the possibilities?

One thing is almost certain: the power of the computer will continue to advance to unimaginable levels. Before 2050, the price of a computer with the computational power of everyone on the planet combined will be less than you paid for your smartphone in 2018. All these issues we discussed, including whether we are machines or not, will become settled science. But more processing power will mean nothing if it doesn't power some kind of state change in the device. A calculator is a powerful device. A stadium filled with a billion calculators is in no way more powerful. Faster computers will be one of two things: something emergent and new, or simply a computer that runs faster.

Many believe the true supercomputer will achieve true superintelligence. In fact, there are even those who are rooting for the machines over humanity, who believe we should build our replacements, and it is only our shortsighted selfishness that believes otherwise. Dr. Hugo de Garis, a retired professor and AI researcher, is one who holds such views. He identifies himself as a Cosmist, and maintains that AGIs are the next phase of evolution and will supplant us and populate the universe. He believes that the fate of one species on one planet is of no real concern compared with the fate of the universe and says, "Humans should not stand in the way of a higher form of evolution. These machines are god-like. It is human destiny to create them."

When Might We Get AGI?

Assuming we can build an AGI, when might we build one? No one knows. As the physicist Niels Bohr said, "Prediction is very difficult, especially if it's about the future." AGI is no exception. As I mentioned earlier, dates from knowledgeable people range from five to five hundred years, which is about as useful as your dry cleaner saying your shirt will be ready in anywhere from five to five hundred days. But this uncertainty reflects a simple fact: it is hard to predict something that has never happened in the history of the world, especially if that thing is something we don't understand or know how to replicate. Other than that, it is all easy.

That's all well and good, but why is there such a huge disparity of guesses—two orders of magnitude, to be exact? A few reasons: First, there is disagreement on how complex intelligence is. Some people believe we are just a couple of big breakthroughs away from a general learner, something you can just point toward the Internet and say, "Go learn all of that." Others think intelligence is so complex that we will have to solve hundreds of devilishly hard problems one at a time. Both positions point to biology as a proof point: the relatively small-size "code" in human DNA suggests that generalized intelligence can be built with a little code. But the fact that the human brain has hundreds of specialized abilities located in its various regions suggests our intelligence is, in computer terms, a giant kludge, an unholy mess of spaghetti code.

Additionally, there is a difference of opinion on whether or not we are on our way already to building an AGI. Some people see an AGI as an evolutionary advance from the science of narrow AI today. Others believe that today's AI is a different thing entirely from general intelligence, and we don't know when the needed breakthrough will happen. I often ask the guests on my AI podcast, *Voices in AI*, which of these camps they are in, and I find the experts to be pretty evenly split on this question. People who think an AGI will come from the same basic techniques we use today often believe we will have an AGI in twenty or thirty years. The other group thinks it will be much longer.

Also, the assumption of most parties is that there will be a moment when we get enough of a spark for a proto-AGI to take over its own development, which will allow it to advance quickly into a full-blown AGI, and then perhaps even to a superintelligence. Some think that this spark will happen relatively soon as we teach the machines basic principles for gathering data and drawing inferences from it. Others think that we are going to have to build a great deal more of the AGI before it can take off on its own.

Finally, some people hold human intelligence in higher regard than others do. I chat with many AI practitioners who don't think human creativity, as an example, is particularly mysterious. I'm not saying these folks are down on humans; rather, they generally look at human ability as being straightforward. The professor and author Pedro Domingos, a giant in the field, told me, "Automating creativity is actually not that hard," and rattled off example after example of computers writing music, news stories, and the like. Meanwhile, another group sees human intelligence as a vast array of skills carefully selected for across millions of years, and suspect that duplicating that is a monumental task.

So the people who think intelligence is straightforward, creativity is not all that special, and we are already on our way to building an AGI, which will soon take over its own development, are those who speculate that we will have an AGI within a decade. Conversely, people who think we are five hundred years away from an AGI view intelligence as inherently difficult, see human mental ability as extraordinary, and believe we haven't even begun to go down the correct path.

Those who actually write code and work in the field generally believe that AGI is further away. While almost universally believing we will one day make an AGI, they are down in the trenches building narrow AI to do the simplest of tasks. They are, as a general rule, annoyed by those who make broad sweeping statements about an AGI being imminent. One practitioner complained to me that every time a cultural luminary makes some dramatic proclamation about an AGI that it "sucks the air out of the room" and is a major distraction for a couple of weeks.

How would we know if we had created an AGI? Of course, maybe we already have, and it has enough sense to keep its mouth shut for fear of everyone's freaking out. Absent that, some offer up the well-known Turing test as the first hurdle an AGI candidate would have to clear. Turing, whom we discussed in chapter 4, was an early computer pioneer. A genius by any definition of the word, he was instrumental in cracking the Nazis' Enigma code, which is said to have shortened World War II in Europe by four years. Regarded today as the father of AI, Turing, in a 1950 paper, posed the question of "can machines think?" and suggested a thinking test we now call the Turing test. There are varying versions of it, but here are the basics: You are in a room alone. There are two computer terminals. You can type questions on them. On one, the questions will be answered by a computer. On the other one, by a person. You have five minutes. Try to figure out which is which. If a machine can trick you into picking it 30 percent of the time, Turing argued that you must say that the machine is thinking, because it is able to duplicate the capabilities of a person who is thinking. It doesn't matter, in Turing's view, that the machine is doing its thinking differently than a human does. He predicted machines would accomplish this by the year 2000.

The 30 percent number is interesting. Why isn't it 50 percent? Shouldn't the machine be indistinguishable from a person to pass the test? Not necessarily. The point is to see if the machine thinks, not if it thinks as well as a human. The interesting thing is that if computers get good enough at the test to get picked 51 percent of the time or more, then the unsettling conclusion is that they have become better at being, or at least seeming, human than we are.

Critics of the test say it tests only whether a computer program can convincingly emulate human language. They would say this is not AGI, which needs to be as intelligent as a human, which speech alone doesn't demonstrate.

Regardless of your thoughts on what the Turing test actually proves, it is still quite useful in the sense that teaching a computer the infinitude of nuance involved in using language, along with enough context to decipher meaning, is a really hard problem. Solving it has real benefits, since it would mean that we could use conversation as our interface to machines. We could chat with a computer as casually as we do with each other.

The surprising thing is how far away we are from creating something that can pass the Turing test. If you read the transcripts from contests in which programmers actually conduct Turing tests, you can generally tell with the first question whether the respondent is a computer or a person. Computers aren't very good yet. And these are events at which the rules have been stacked to favor the machines, by banning certain topics and limiting the kinds of questions. The candidate AIs choke on questions like "What is bigger? A nickel or the sun?"

Their poor performance so far is understandable. We've had 100,000 years to work on language, and in that time we have become pretty good at it, but Turing's simple test shows just how difficult it is to replicate our ability with language, which is one of many requirements of an AGI. If a machine cannot understand the simplest natural-language question, think about how far it must be from emulating all human ability. Right now these systems don't even do a good job of understanding the question, let alone answering it. In the example above, is a nickel a coin or the metal? Since it is "a nickel" and not just "nickel," we understand, but that is tricky for the bot. And is it "the sun" or "the son"? Humans know. But imagine the depth of knowledge needed to answer the following question:

> *Dr. Smith is dining at his favorite restaurant and gets an emergency call. Dashing out, he neglects to settle up with the waitstaff. Is management likely to prosecute him?*

A human can figure that out easily. A doctor gets an emergency call and has to rush to the hospital, forgetting to pay his bill. Since it is his favorite restaurant, he is probably known there, and it is therefore unlikely that the management will call the police. How does a computer extract all that nuance and inference? It is quite difficult. Or consider this question:

> *What would be a good plot for a story about a time-traveling princess who has a young pet dragon who can't control his fire breathing yet?*

A human could come up with dozens, like "The pet dragon catches a cold and every time he sneezes fire shoots out of this nose," or "The dragon gets lonely so they travel into the future to have a sibling cloned for it."

I think an even more interesting test, however, is what I call the preschooler test. You want to impress me? Build an AGI that can answer all the questions of a four-year-old. Four-year-olds ask, on average, over one hundred questions a day, such as "Are there more hairs or blades of grass in the world?" or "What does your last day as a kid feel like?" or "How did people make the first tools if they didn't have any tools?" or even, "Why does Mr. T pity the fool?"

So while we don't know when an AGI will arrive, be it five or five hundred years, it is worth noting that a number of predictions cluster around the 2030s, which isn't that far off. Those who make these predictions point to the advances we are currently making and the power of computers at that time. But all predictions about the timing of an AGI should be taken with some amount of skepticism given that there isn't consensus even on how to build an AGI.

AGI and Ethics

An AGI presents two distinct ethical challenges. The first is how to make an AGI that *behaves* ethically. Assuming that we are the ones setting its goals, we would want its values to align with ours and act both ethically and humanely. The word "humane" is simply a variant form of "human." The meaning of the word encapsulates us at our best, not at our average. But how can we do that? Technical issues aside, in simple English, how would we teach a machine to act ethically?

Isaac Asimov took an early stab at it in a short story he wrote in 1942, which he later developed into his I, Robot series of books. Asimov coined three laws programmed into every robot designed to ensure that the interests of the robots never collided with that of humans. The laws were:

> *A robot may not injure a human being or, through inaction, allow a human being to come to harm.*
>
> *A robot must obey orders given it by human beings except where such orders would conflict with the First Law.*
>
> *A robot must protect its own existence as long as such protection does not conflict with the First or Second Law.*

While his three laws are a worthy attempt, they only prove the point of how difficult it is to encapsulate complex ethics in a series of simple rules. For example, take the first law. An AGI, driving you down the road, might be forced to either run over a child who just ran into the road or run the car into a tree to avoid hitting the child but potentially killing you. This decision is not only morally challenging, but legally so as well. Expect to see programmers subpoenaed and potentially charged with crimes for the code they write. I can hear it already: "Is the court to understand that you deliberately programmed the car to drive into a tree and kill the passenger?" This is obviously an edge case, unlikely to be a day-to-day occurrence, but there are many other wrinkles that emerge from this seemingly simple rule. What about something that has a chance of hurting a human? A tiny chance? What if the human is about to hurt him- or herself, even accidentally? Can the robot forcibly stop the human? Can "hurt a human" be interpreted as emotional distress as well? If you decide to celebrate an accomplishment with a cigar, should the robot stop you to protect you? What if you resisted? What about if you wanted to eat a second dessert? Doesn't the robot have a duty to intervene?

So if Asimov's rules won't help us, where do we turn next? Perhaps we can code an ethical standard from the ground up, so to speak. In this case, there are three big challenges when it comes to making an ethical AGI. First, humans don't have a shared agreement on what constitutes an ethical standard. In fact, the palette of acceptable ethical standards in the world is quite broad and often contradictory. Second, moral codes are moving targets. Just think about how our views on the proper ways to interact with others have evolved over the last century, even over the last decade. We have no reason to believe our views of right and wrong will not continue to change. The third challenge is that even if you got past all of that, our own individual moral codes are encumbered by a Byzantine set of exceptions, provisos, and special cases that hinder the moral code from being easily instantiated into a computer program. Asimov thought he could do it with three laws, but a person's moral code contains thousands. Consider the following:

Is stealing wrong? *Yes.*

Is stealing from the Nazis in World War II wrong? *Well, no.*

Is stealing bread from a hospital for children in Nazi Germany during World War II wrong? *Oh. Well, then yes.*

Is stealing some of the surplus bread from a hospital for children in Nazi Germany during World War II when you yourself are starving wrong? *Hmm, I guess not.*

As you might guess, this can go on ad infinitum.

A project backed by the Future of Life Institute to work on the problem of instilling ethics in an AGI described the challenge this way: "Some AI systems do generate decisions based on their consequences, but consequences are not all there is to morality. Moral judgments are also affected by rights (such as privacy), roles (such as in families), past actions (such as promises), motives and intentions, and other morally relevant features. These diverse factors have not yet been built into AI systems."

This is a hard problem. As AI goes into more critical areas, such as medicine, the ramifications of the ethical decisions it makes grow ever larger. Make no mistake: the AI will have an ethical standard programmed into it. It may not even be intentional, but any time the AI has to decide between X and Y, it will be applying a value system of some sort, and it almost certainly will have unintended consequences. At present, the value system is that of the programmers making the AI. If we ever made an AGI, the implications of the moral values of that system are dramatically more profound.

The second ethical challenge with an AGI is how to *use it* ethically. Several ways to ethically misuse an AGI pop immediately into my mind, and I don't even possess the diabolical cunning of a criminal mastermind. An AGI would be a powerful tool to use to steal and swindle in new and innovative ways. It is an old dance, as old as civilization; however, those on the side of law and order would also use an AGI to thwart those same criminal enterprises.

An AGI can also be misused in other ways. For instance, states, corporations, or other entities could use it to invade privacy in new and profound ways. For example, it could ingest huge amounts of Internet traffic, effectively seeing what everyone is typing and looking at. It could read everyone's emails. With voice recognition, it could listen not only to every phone call, but also to conversations everywhere near a microphone. Cameras already blanket the world, and face recognition is coming into its own. Researchers at Oxford University and Google Deep-Mind have made great strides forward in lip reading, which could be combined with the cameras. The result of all of this? A machine that would be effectively both omnipresent and omniscient.

There is nothing surprising about an agency of some sort being able to gather all that data. That's old news. Privacy was always protected due to the colossal size of the data sets. We were each lost in the crowd. No human could make sense of it, and no computer could sort it all out. But imagine an AGI, which could understand all that speech, recognize all those faces, track all that movement. It could effectively create a master file of everyone, right down to the most obscure details. It could then connect all the dots, figure out all the connections, surmise every plan, calculate the possibility of everyone doing anything, and generally remove all privacy except, hopefully, that of private thoughts.

This is not foil-hat-central talk. We are speedily developing these tools for non-nefarious reasons. Almost everything you read about big data, whether it is in regards to medicine or meteorology or anything else, implies that it requires these sorts of tools. The same technology capable of identifying connections in symptoms of cancer survivors is the same technology capable of identifying connections in political affiliations. Just different data. And by the way, none of this really requires the superintelligent self-learning power of an AGI. We know how to do all these things now. While the possibility of creating an AGI is still an open question, we are steadily improving these narrow AI technologies. Over time, we will get better at it, and processors will get faster, and more of the world's life will be lived online. All it would take would be a shift in the laws of a nation to transform the Internet from an information tool into a surveillance tool. There is no doubt it has already happened in some parts of the world.

How do we keep this from happening? The simplest way is just not to do it. The problem is that it will come in bits and pieces, none of which alone is worth a protest march. It will then accelerate quickly in the interest of safety and security. In the end, however, the surveillance state will come only if people clamor for it, if they demand it. Freedom often dies by popular acclamation.

Another potential to misuse an AGI may not seem as bad, but it bears thoughtful consideration. It relates to the emotional and potentially dehumanizing aspects of AGI. Earlier, I told the story of the early AI pioneer Joseph Weizenbaum, who later turned against the idea of AI when he saw how people interacted in an emotional way with ELIZA, his simple AI therapist. He wrote a landmark 1976 book called *Computer Power and Human Reason* in which he maintains that any job that requires true empathy, such as eldercare aide, soldier, or even customer service representative, should never be done by a computer. He believed that extensive interactions with machines emulating human empathy would make us all feel more isolated and devalued.

His point is well taken. Maybe we have all had the frustrating experience of being caught in some infuriating bureaucratic system, trying to get something done, and having uncaring humans tell us we were missing some form and instructing us to go stand in yet another line. The total lack of empathy is isolating, and to imagine that life consists of more and more interactions that merit empathy but are instead substituted by an AI or AGI programmed to merely pretend to be empathetic is concerning.

Further, Weizenbaum makes a distinction between deciding and choosing, and suggests that computers should do only the former. Deciding is computational, like deciding which is the shortest route to work. Yet he believes that machines should never choose. A human might choose to run his car into a tree to avoid a child, but that is a human prerogative, and Weizenbaum felt deeply that delegating this to machines in no way ennobled the machine but rather debased the human. Weizenbaum felt that our temptation to delegate core parts of humanity to computers indicated "atrophy of the human spirit that comes from thinking of ourselves as computers," a phrase that harkens back to our foundational question about what exactly we are.

Additionally, if we do make lifelike robots that we program to do all of the dull, dirty, and dangerous tasks we don't want to do, could this have an unintended corrosive effect on how we regard other humans? Imagine a lifelike robot with an AGI that helps Granny out around the house, allowing her to live independently. Further, say that over time we make robots more and more humanlike in appearance, displaying facial emotions and so forth. Not indistinguishable from humans, just human-oid. How would we treat this entity? On the one hand, the AGI is like a human, full of simulated wit, affection, and empathy, so we might treat it with a reflexive kind of courtesy. On the other hand, it is just a bunch of wires and gizmos and we should have no more regard for it than the plunger it uses to unclog a toilet. In that case, we may have no regard for it at all. This conflicted view of the AGI would be difficult to maintain over time. Might this disposable humanoid robot inadvertently cause us to value human life less?

Researchers in Japan performed a study that suggests this might be the case. A robot, Robovie II, was put in a mall. It had been programmed to politely ask a person blocking its path to step aside. If the human didn't, the robot went around the person. It turned out that children were often inclined to give the robot a hard time. They would deliber-ately block the robot's path. But that's not all. Often kids' behavior would escalate from there. They grew violent, kicking, hitting, and cursing at Robovie. A couple of times, they hit the robot with a bottle.

The researchers were interested in how to program the robot to avoid these situations. They made a model of who was likely to harm the robot, and it turned out the likeliest candidates were groups of kids when no adults were around. So they programmed the robot to flee those situa-tions and race toward a tall person, presumably an adult.

Robovie doesn't look like a person. However, the disturbing finding was that when the researchers asked the children who had bullied the robot if they thought it was humanlike, 74 percent said yes, with only 13 percent agreeing that Robovie was machinelike. Fifty percent said they thought that their actions were stressful or painful for Robovie.

So the question is whether or not this pattern of behavior becomes habitual as the devices get more sophisticated, and whether that behavior then spills over into interactions with other people. One can imagine that bullying a robot could possibly have a numbing effect on humans.

Are We Living in a Simulation?

Is it possible that all of this has already happened, and someone has built an AGI that can house simulations that either are, or believe themselves to be, conscious? Is it therefore possible that *we* are in fact entities living in a simulation? If so, the question of why there is all the pain, misery, disease, and suffering that exists in the world takes on a new dimension. Suffering is certainly real, at least to those who suffer. This question of suffering in a simulation is touched on but then dodged in *The Matrix* when Agent Smith explains that at first the Matrix "was designed to be a perfect human world, where none suffered, where everyone would be happy." But he goes on to explain that it was a disaster because our brains wouldn't accept that as possible. But this is a Hollywood answer. There are plenty of actual societies with relatively little pain and suffering. So if our reality is a simulation, it is seems neither intentionally benevolent nor cruel.

The idea that everything around us is an illusion is an ancient mystical idea. The simulation hypothesis is that same idea dressed in modern clothes and carrying an iPhone. Neil deGrasse Tyson says that the likelihood that it is true "may be very high." Other adherents include a wide range of intellectuals, science fiction authors, and Silicon Valley types, including Elon Musk.

Broadly speaking, there are two arguments in favor of the simulation view:

First is the statistical probability argument. This says, "At some point in time, we will be able to develop completely realistic simulations of our universe inside a computer, and inhabit them in digital forms. Given the age of the universe, it is reasonable to assume that other civilizations have already done so. Once they are created even one time, an infinitude of copies of the simulation could be made. Given the size of the universe, a nearly infinite number of species have probably already developed that simulation technology, each of them capable of being copied millions of times. Therefore it is highly likely that there is one "real" universe and a trillion digital ones. Statistically speaking then, we almost certainly live in one of the trillion.

I find this line of reasoning unconvincing. It is like one of those statements in which every individual step is true but the conclusion is misleading, like "wearing seat belts increases cancer rates" or "virtually all murders occur within ten hours of the murderer eating." I think the setup is disingenuous, and in fact, we have no reason to believe *any* of the statements in it are true. We have zero reason to believe a simulated human would have subjective experience, any more than we would expect that Pac-Man really feels pain when being hit by a ghost. We have no reason to believe an alien species would make a simulation, no reason to believe that a synthetic universe, if developed, would be copied ad infinitum, no reason to believe that whoever made the simulation would want to trick us into thinking we are real.

The second argument in favor of the simulation hypothesis is that we can actually detect a glitch in the matrix. It goes like this: Any simulation must be inherently finite, because the computer running it is finite. As such, the model would have boundaries, likely expressed in the form of limits, which we understand as the physical laws that bind our universe. Further, a simulation wouldn't be completely built out. Think of the billions of galaxies out there, each of which is composed of billions of stars, each of which consists of an octodecillion atoms (a one followed by fifty-seven zeros). Each of those atoms has its own speed and trajectory. It is unlikely that someone building such a simulation would use the processing power needed to keep track of the location of each of those atoms when, from our vantage point, those galaxies are but pinpricks in the night sky. Some scientists maintain that we can detect evidence of such limits and computational shortcuts.

There is a vigorous discussion of this possibility online, and a good place to start if you are inclined to take the red pill is a paper called "Constraints on the Universe as a Numerical Simulation" by Silas R. Beane and his team at the University of Bonn.

But I personally avoid drinking too deeply from that well. Speculation as to whether everything we know and everyone we love exists only in the thumb drive of an adolescent alien seems, oddly enough, irrelevant. Whether we live in a simulation or not, we know of only one universe, and only one planet within it that harbors life. There are many things that we as a species have yet to do, including feeding the hungry, curing disease, and making a world where all people can achieve their maximum potential. Simulation or reality, pain and suffering are real things to those who experience them.

Part Four

COMPUTER
CONSCIOUSNESS

THE STORY OF JOHN FRUM

In the Pacific, near Australia, is an area known as Melanesia, which consists of four countries: Vanuatu, the Solomon Islands, Papua New Guinea, and Fiji, as well as some smaller islands. These nations were in the crossroads of the Pacific theater of World War II. As a result of their interaction with the American military, a strange sociological phenomenon occurred that came to be called cargo cults.

The indigenous people of these islands would see an American force land, clear some ground for a runway, and build an observation tower. Then the islanders would gaze with awe as planes would arrive from the sky, land on the runways, and offload enormous amounts of cargo. Often the military would share with the islanders some of the bounty, including canned food and manufactured goods.

Thus were born the cargo cults. The local people would clear their own runways and erect their own towers, but from bamboo. Lacking a radio, they might fashion a box that resembled one out of coconuts. They didn't have lights to guide planes in, so they would plant bamboo along the runway. Using wood for guns, they would perform military drills the way they had seen the Americans do it, often in costumes designed to look like US military uniforms.

Occasionally they would even build full-size planes out of straw in the hopes they would attract other planes. They did everything the Americans did. But oddly, the planes never landed and the cargo never arrived.

Even today, a group in Vanuatu worships John Frum, an idealized American serviceman from World War II. Each February 15, they hold a parade honoring him and the belief that he will return some day with cargo for everyone.

16

Sentience

The question we are about to address is whether a conscious computer is something that can be built, or if those who try to create one are just members of a technological cargo cult, deluded in their belief that if they build a machine a certain way, the planes will land and the machine will be conscious.

We are going to address the consciousness question head-on, but to do so, we first need to explore two concepts, sentience and free will. A full understanding of these will go a long way to determining whether machines can become conscious, for while they are not synonymous with consciousness, they do share some essential traits with it. Let's begin with sentience.

Sentience is a word that is regularly misused. In science fiction stories, it is usually used to mean "intelligent," as in "The beings on Rigel 7 have begun forming cities and are clearly sentient." But it doesn't actually mean that. Sentient means the ability to feel, or "sense," things. The word that should be used in the Rigel 7 example is "sapient," which means "intelligent." That is why we are *Homo sapiens* ("intelligent people"), not *Homo sentients*.

Although it doesn't mean "intelligence," sentience is still an important concept to us. We care if animals, for instance, are sentient. Can they feel pain? If they can, then we are mindful of that in the ways we treat them. The bacteria I wipe out with antibiotics aren't sentient. The mosquito I swat isn't either. But dogs feel pain, as do cows and apes. Thus we have laws against animal cruelty, but we don't apply them to jellyfish and tapeworms. There is no broad agreement as to where sentience begins in the animal kingdom, but we do know with some confidence about the extreme ends.

Can computers be sentient? Can a machine theoretically ever *feel* something such as pain? The answer matters a great deal to the question of computer consciousness. Nothing can be conscious that cannot have experiences, so sentience is by necessity a prerequisite for consciousness.

When I was a boy, I had a Weimaraner named Misty. Once when my buddy Steve and I were playing with her, she jumped over a water faucet and in a freak accident ripped open her front leg. Her cries sure sounded like pain to me.

In 1998, Tiger Electronics released a doll called a Furby, a robot toy that was immensely popular, selling forty million units in three years. It gradually "learned" human language over time and had a few rudimentary sensors built into it. When the Furby was held upside down, it was programmed to say "I'm scared" in a plaintive voice. Was it scared? I doubt anyone would think so.

What is the difference in these two cases? What type of thing is pain?

If you plug a sensor that can detect heat into a computer, then program the computer to say "Ouch, ouch" when a match is held to the sensor, is the computer feeling pain? If your answer is no, then how are the computer's cries of pain different from those of my dog Misty?

Recall our foundational question about the composition of the universe. Monism or dualism? Monism, as you might recall, is the position that everything in the universe is made of atoms and governed by physics. That's why it is often called materialism or physicalism. If you are monist, the abstract idea of pain actually poses a bit of a problem. The monist might say, for instance, that "pain is pure physics. Hit your thumb with a hammer and you will come to have a deep appreciation of that fact."

But in this example, what exactly is pain? The monist might offer the dictionary definition of pain as "an unpleasant sensation that occurs in the brain that is often triggered by disease or damage." But this definition just kicks the can down the road. The word "unpleasant" just begs the question of what pleasant is, which is the same kind of question as what pain is. "Sensation" is equally problematic because the word itself is something we are still trying to get our arms around. What does it mean to sense something? Further, locating pain in the brain would imply that an alien life-form or a strange new creature or an intelligent machine that doesn't have a literal brain cannot feel pain. An alien may have painted the alien version of the *Mona Lisa*, but if it stubbed its toe and leaped about saying how much it hurt, the monist adhering to that definition of pain would have to say that it wasn't in fact feeling pain given that it didn't have a brain. This view is clearly unsatisfactory. What sort of definition of pain could a monist choose that includes humans and animals, excludes Furbies and bacteria, and leaves the door open for trees and computers? It is hard to say.

If you are a dualist, you have a different problem coming to grips with pain. You are fine with "pain" being an abstract thing, a nonmaterial thing that exists. You don't even have to localize it in the brain. You are fine with an understanding of pain that could apply to a person or an animal, exclude Furbies, and leave the door open, hypothetically, to plants and machines. Pain is not a physical thing, but a mental one. That's what dualism is all about. But you still haven't gotten around the main argument monists make against dualism, which is the question of how the mental and physical worlds can interact. If pain is a nonmaterial thing that exists outside the physical world, then, as the monist would ask, "Why does hitting your material thumb with a material hammer cause this nonmaterial sensation?"

Regardless of which position one takes, there is one commonality between both points of view. For pain to exist, something has to feel the pain. Whether you are a monist or dualist, there must be an "I" that senses the pain. My dog Misty appeared to me to have an "I" that felt pain. This is not to say that she was conscious; we will get to that later. But she seemed to have a self that could experience pain. This brings us back to our foundational question about what your "self" is.

If you believe your "self" to be a trick of the brain, an illusion of the many parts of the brain working in tandem, then it seems reasonable that we can build a computer that can perform a similar trick. That would be a sentient computer. That isn't a conscious computer; rather, it's simply one that can feel something. If you hold the view that your self is an emergent property of your brain, then we have to keep the question open, because emergence is not understood well enough to know if it can be replicated mechanically in a predictable way. And finally, if your idea of "self" is something noncorporeal such as a soul, spirit, or life force, then a computer that feels is probably out of the question unless you believe that force can inhabit man-made objects. While many people do believe that places can be haunted by ghosts, it is difficult to see how that might be manufactured at scale in a computer factory.

What life-forms can feel pain? As noted earlier, this is an open question, and one that is hotly debated. Let's work our way from humans to plants and see what we can conclude.

You know, of course, that you feel pain. You feel it. You are pretty confident you know when other humans feel pain because they self-report it, and unless they are lying, they are the authorities on whether they are in pain. We infer mammals are in pain because they exhibit many of the same signs as humans do and appear to be affected by the same sorts of things that cause pain in us. If an ape hits his thumb with a hammer and howls, we conclude that it hurt him too. If one ever utters an expletive in sign language, we would take that as additional proof. However, there is no consensus on the question of mammalian pain. Veterinarians trained in the United States prior to 1990 were taught to ignore animal pain, as it was believed animals did not actually feel anything. This topic still matters in a practical sense because it speaks to medical and product testing on animals, and to the use of animals in scientific research. It should also be worth noting that until the latter part of the twentieth century, the consensus medical opinion was that babies didn't feel pain either and were therefore operated on without anesthesia.

The fact that there isn't universal agreement on whether mammals feel pain is worth looking at. How can people maintain that viewpoint? How could it have been a mainstream opinion so recently? The argument goes something like this: "Just because a dog recoils when poked with a needle doesn't mean anything more than the fact that a single-celled amoeba does as well. It is simply a response programmed into the animal's DNA. We are projecting the feeling of pain onto the dog because we would feel pain under similar circumstances."

The argument that mammals don't feel pain certainly sounds to my ear like one made from convenience. It allows one to avoid all kinds of ethical problems relating to how we treat and use animals. I doubt anyone could convince me that my dog Misty didn't feel pain, but in relating the story I deliberately used the phrase that she "appeared to" because in the strictest sense I cannot "know" that she did.

Moving beyond mammalian pain, we should find it no surprise that there is even less consensus with respect to fish and invertebrates. You can't gauge their pain by their behavior, for an animal might feel excruciating pain but hide it at all costs lest it become a target for prey. There is some confidence that insects do not feel pain, which is why you sometimes see one walking around as if it had not a care in the world in spite of having body parts missing. Coral, about as simple a life-form as there is, also don't seem to feel pain, as they react to gentle, nonharmful touch in the same manner as a violent poke.

How about plants? You may scoff, but there is some interesting research that suggests that more is going on there than meets the eye. The idea that plants can feel sprang up (or more accurately, was resurrected, for it is an ancient belief) in *The Secret Lives of Plants*, a 1973 book by Peter Tompkins and Christopher Bird that maintained that plants have a wide range of emotions. The book also referred to a CIA report of a man hooking up a polygraph to a plant and the instrument responding when he just *thought* about burning a leaf, suggesting plants are telepathic as well.

These sorts of experiments have been difficult to reproduce, and botanists are dubious, but the topic is serious enough to have merited a feature in *The New Yorker* in 2013 in which the idea is considered across ten thousand words of copy, in which plenty of credible scientists were happy to make the case, or at least consider the argument.

I bring the issue of plant pain up not because I think plants are sentient, but to point out that *we wouldn't know* if they felt pain or not, because such pain would be alien to our ideas of what pain feels like. This is not a merely academic question, for if we can't know how plants, with which we share a majority of our DNA, feel or don't feel pain, how would we know if a computer did? Even if the computer said, "Ouch, that really hurts," isn't that just a product of its programming? Even if it programmed itself, how would you know whether it felt anything at all? One could imagine an AI tasked with a problem concluding that if it claims turning the computer off hurts it, it might be able to get the humans to leave it on and it can better achieve its task. It could arrive at that, a mathematical conclusion, even though it feels nothing at all.

The fact there isn't widespread consensus on the various issues around sentience we have just explored shouldn't surprise anyone. After all, questions like these kept Socrates awake at night. But hopefully you, as a reader, have some personal clarity on them. Answer the following questions according to your beliefs, keeping track of the number of times you answer yes:

(1) Do dogs feel pain?
(2) Is it possible that invertebrates like insects and spiders feel pain, albeit differently than we do?

(3) Is it possible that plants feel pain?

(4) Is it possible that a sophisticated alien without a nervous system could feel pain a completely different way than humans do?

(5) Is it possible that *anything* inorganic (that is, that doesn't contain carbon) could feel pain? This could include crystals, thunderstorms, an alien made of water and dirt, or anything else inorganic.

The more times you answered yes, the more likely you are to believe that machines can be sentient.

17

Free Will

The next topic to discuss before we address the question of computers becoming conscious is free will. Why is this germane? In your daily life, you have experiences, which are the basis of consciousness, and you make choices, which is free will. You have a conscious experience of tasting pineapple, then you exercise free will to take another bite. They feel like opposite sides of the same coin, and it feels like the same you that is experiencing the pineapple is exercising free will.

If free will turns out to be completely mechanistic, effectively an illusion, it will suggest, though not prove, that consciousness also fits that mold. If, however, free will is something that exists outside materialistic cause and effect, then consciousness might also as well. Further, if free will does exist, then two questions come to mind: "Where does it come from?" and "Will computers have it?"

We understand the material world through two different sets of physical laws, one for large objects and one for small objects, neither of which seems to allow for free will. The one for large objects is Newtonian

215

physics, in which every action has a cause. Every bit of your life, right down to stubbing your toe on the bed frame this morning, was the inevitable outcome of a series of events that dates back to the big bang. The set of laws for small objects is the quantum view, in which there are actions with no causes, and those actions are completely random. The decaying of a radioactive substance is often used as an example. It decays at a certain rate, but which individual particle decays is entirely and truly random in the most fundamental sense of the word. So it would seem you are, at your core, either a robot or a random-acting lunatic.

On the one hand, we have a thing that feels like free will. On the other, we don't have a great way to explain how it is possible. Because of this, people break into two camps: the libertarians (not the same libertarians as the political ideology), who believe in free will, and the determinists, who don't. The core question for our purposes is whether or not your brain is a causal system. In other words, is your brain like a giant clockwork mechanism? You wind it up, and it runs its rigid, unchanging program. No one thinks a clock has free will. It just does exactly what it has been engineered to do with absolutely no say in the matter. Is that what you are as well? Is the belief that you are making choices just a delusion? Generally speaking, dualists believe in free will, whereas monists don't.

How does neuroscience regard free will? Leon Gmeindl and Yu-Chin Chiu of Johns Hopkins University did an interesting experiment in which they placed participants in MRI machines. The subjects could see a monitor that had colorful letters and numbers scrolling by on the right and the left, with nothing in the middle. They were asked to observe one side, switch their mental focus and look at the other side, and then, at some point, switch back. They weren't given instructions on when to do all that. The idea was that whatever part of the brain was active the instant before they switched focus must be the place where "free will" is located in the brain, or at least the free will to change mental focus. And sure enough, Gmeindl and Chiu found consistent places in the brain where activity occurred immediately before the shift. So is that where so-called free will lives in the brain? If so, it sure looks like plain ol' deterministic brain activity. There is nothing special going on at all. The brain is just doing its thing. However, it isn't that simple. The plot, as they say, thickens.

In a seminal paper published in 1999, the psychologists Dan Wegner and Thalia Wheatley proposed a revolutionary idea. Instead of the traditional order of the sequence—a person decides to do something and then it happens—they maintained that things in the brain actually run backward from that. First, the theory goes, you do something, then you tell yourself later that you decided to do it. Can that really be true? Life sure doesn't feel that way. What could be going on here? Good question.

Adam Bear and Paul Bloom, both psychologists at Yale, have an answer: Different parts of your brain, each specialized, tell your body to do different things, such as (per our example above) view the other side of a screen. Then after you do start looking at the other side, your conscious mind notices this and quickly guesses why you switched. It then says to you, "You should move your eyes now and see if the numbers on the other side are the same as this side." Then—and here is where your conscious mind gets sneaky—it rewrites your memories to make you think you decided *first* to switch sides, *before* you did so. Bear and Bloom did some pretty compelling experiments to demonstrate this, experiments that can be reproduced with no more equipment than a PC and a simple program. So, in their view, your feeling of free will, that feeling that you are deciding things, comes a few milliseconds *after* you actually do the things. Let me say that again: your decision to do something seems to occur after you do the thing. It is as if your conscious mind is forever running after your body yelling, "That was my idea! That was my idea!"

Why would our brain play this trick on us? Why would we be made to think there is an executive "I" that calls the shots? One theory is that a belief in free will is beneficial. The psychologists Kathleen Vohs and Jonathan Schooler published a report in 2008 that seems to lend credence to this theory. Two groups of students, one of which had just been read a text about how free will doesn't really exist, were given a math test.

The group that had been read the text cheated more than the group that didn't, suggesting that a belief in free will is important for encouraging moral choices in society. Additionally, perhaps because we think free will exists, we are more willing to punish criminals and hold people accountable for their actions, all of which goes to build a civilized society.

Although the experiments are quite compelling, all of this is by no means proven. We don't even understand how memories are encoded in the brain. We refer to what we can measure in the brain by the most nebulous of terms, "activity." It is like someone describing all that goes on in New York City as "movement." At present, as far as neuroscience goes, all we can see is that things are happening, so a dose of humility is still in order here.

Having said that, let's entertain the notion that it is all true, that your brain really does make decisions on its own that you later try to justify. A monist, or a person who believes humans are machines, would see this as an additional proof of the monist position. The monist may see so-called free will as complex or chaotic or illusory, but whatever it is, it is happening in the brain and is governed by physical laws. Your brain is like the clock, ticking away, and "you" have no say in what it does.

A dualist, as well as a person who sees humans in nonmechanistic terms, has no problems reconciling her belief in free will with the experiments we just discussed. While dualists believe the brain undoubtedly controls the body, they see the salient question as *why* the person chose to look at the other side of the screen at just that instant. What triggered that brain activity that was observed? And for every answer offered, they would in turn ask, "And what caused that?" with the infuriating persistence of a four-year-old. Eventually, in this view, you either get back to the big bang or you spot free will.

The monist would probably roll her eyes during the dualist's monologue, replying in an exasperated voice that we have absolutely no reason to believe free will is anything other than a deterministic mental process. The dualist would feel compelled to reply, "Except our own daily experience of choosing things."

The great Samuel Johnson captured the conflict quite well centuries ago. When asked if he believed humans have free will, he replied that all theory holds that we do not and all experience holds that we do. It is certainly true that when we look inward, we don't feel ourselves operating with the same mechanical precision of a clock or the orbit of a planet. What it feels like is that we have vibrancy, will, intention, drive, and ambition. These could all be illusions, but virtually everyone on both sides of the issue acknowledges it feels like we have free will.

As with several of the questions we have looked at so far, there is no easy resolution. If free will does exist, one has to grapple with where it comes from and how it cannot exist in either the cause-and-effect Newtonian framework or in the randomness of quantum physics. And if it does exist outside those worlds, what governs it? Where does it come from? And how does it communicate with the very physical brain? Only understanding how that happens would allow someone to know if a computer could have free will.

Those who deny free will are left in an equally difficult place. We all know what it feels like when your leg kicks up as the doctor tests your reflexes. There is no feeling of free will in the action. Your leg just kicks. But deciding what flavor of ice cream to order has a distinctly different feel to it. What is that feeling? Why doesn't all of our existence feel like that leg kick?

18

Consciousness

What is consciousness? It is often said that no one knows what consciousness is. This is not true. There is broad agreement on *what* it is. The mystery is how it comes about. So what is it? It is that feeling of subjective experience, of all your first-person sensations. You can feel the warmth of a fire burning in a fireplace, but a thermometer can only measure temperature. The difference between those two things is consciousness. It is, simply put, the experience of being "you." It is the thing that makes life worth living, because without it, you are just an emotionless zombie going through life, feeling no love, experiencing no joy. Consciousness really is the single most important aspect of our existence.

Consider this: We understand the "brain" of a simple robot. It walks until it runs into a wall, then changes directions, and tries again. It may exhibit complex behavior, but we know "no one is home," so to speak. There is no "self" within it experiencing frustration. Why aren't we like that? How could simple physical processes in our brain enable us to *experience* things? How can matter feel frustration? Can a rock fall in love? Of course not, but then why can we? We are made of the same exact stuff as that rock, just elements on the periodic table. How can a mass of neurons, regardless of how many there are, become aware? How can mere matter have first-person perspective?

Imagine, for a moment, that aliens kidnapped you and, so no one would notice, replaced you with a robot that looked exactly like you. Further, imagine this robot was programmed to go through the day doing all the things that you would have done. But this robot is just a glorified toaster. It doesn't think anything. It doesn't know what it is like to feel anything. It doesn't know the feeling of loving your spouse or getting annoyed by traffic or wistfully daydreaming. It doesn't know the sensation of recalling a funny joke and chuckling. It doesn't know the experience of contemplating death or wondering what comes after it. It has no idea what ambition feels like. It can never enjoy the refreshing coolness of a glass of cold lemonade on a hot day. It doesn't know what it is like to hear a song and like it. It doesn't ever experience the pleasure of the smell of freshly baked bread or laundry just pulled from the dryer. It doesn't know what heartache feels like. It knows neither the feeling one gets at the height of ecstasy nor at the depths of regret. It doesn't know what it is like to feel brave or cowardly. This robot is just a bunch of wires and motors designed to act like you. To emulate you. So the difference between that robot and the real you is consciousness. But here's the thing: the robot might be able to pull it off. If it smiles when you would smile and writes the memo you would have written, who would know? Yet for some unknown reason, we aren't robots, but something wonderfully more.

Consciousness is different from sentience. A simple life-form might exhibit all the indications of being in pain when injured but have no running mental conversation going on in its brain along the lines of "What in the world did I do to deserve this?" Sentience is sometimes thought of as the lowest order of consciousness, for in a sentient creature, there is at least an "I" experiencing something. Those who identify themselves as machines or animals in our foundational questions will more likely see consciousness along a continuum, in which we simply have more, potentially much more, of it than animals. Those who believe humans are different in a fundamental way from animals will likely see human consciousness as the "something different."

Consciousness is also different from intelligence. Intelligence is about reasoning; consciousness is that feeling of experiencing. For instance, you may have had the experience while you are driving of having your mind wander. You are lost in thought, but somehow you manage to stop at red lights, merge into traffic, and so on. Then suddenly you snap out of it and think, "Oh my gosh. I have no memory of driving this far." That would be a close approximation of being intelligent but not conscious. The difference between those two states is what all the fuss is about.

We are a long way away from understanding what causes consciousness. Why do I say this? It is a phenomenon we have no way to measure. There is no agreement on why we have it or why we need it. We don't even know what the answer to the question "How does consciousness come about?" would look like. How do we answer a question that we don't even know how to pose?

Why did consciousness come about? We don't know that either. You probably drove the car just as well when you were zoned out in the earlier example, so why do we need it? There are plenty of ideas here, such as that consciousness is a mechanism to allow us to change focus easily, but they are all speculation. Perhaps consciousness is wrapped up in our ability to run different scenarios in our heads. A snake can act only like a snake, a possum only like a possum. But humans can watch animals and each other and learn from them, and then imagine multiple courses of action in a situation. That skill is not consciousness, but consciousness might be a tagalong feature, a requirement for this capability. Maybe we need that running voice in our heads to say, "Well, if I climb up to that part of the mountain where no one ever goes, maybe there are plentiful berries on the bushes up there. And wouldn't they taste good!" But this too is conjecture.

Another theory suggests that consciousness came about because of our competitive nature as a species. Competing with someone is easier if you can get inside his head and know what he might be thinking. So that whole process of "putting yourself in someone else's shoes" developed into consciousness. Others suggest the exact opposite is the case, that consciousness came about because we cooperate with each other. To work in unison the way we do requires more than just language. It involves an understanding of the larger goal and your part in it. After all, consciousness enables us to signal each other better. "If I point at that mammoth, Og will look over there and see it charging him." It is worth noting that dogs seem to be the only animal that will look in a direction a human points, suggesting that in the process of domestication we selected for this unusual trait, probably unintentionally.

The philosopher Daniel C. Dennett suggests consciousness might have come about due to the tension between competition and cooperation. Deciding whether to cooperate or compete with someone might be the seed of consciousness. He says, "You can't have language without the possibility of using it to fool people. But also you can't have language without the capacity for cooperation. When you put the two together, consciousness is right there waiting in the wings."

When did it come about? As you have probably guessed, we don't know that either. There is one speculative theory, however, that is worth mentioning. Julian Jaynes wrote a fascinating book in 1976 called *The Origin of Consciousness in the Breakdown of the Bicameral Mind*. Jaynes believed that as recently as three thousand years ago, humans weren't conscious. Instead of being an integrated whole, the two halves of our brain each behaved somewhat independently. One was the commanding half, which would tell the other half what to do. This phenomenon was experienced by the person as a commanding voice. We were all, in effect, schizophrenics. Just as some schizophrenics experience "command hallucinations," our entire existence was governed by commands we received from our right brain, which was reasoning based on our experiences.

Such a brain, according to Jaynes, would not have subjective consciousness, nor would it have the ability to introspect. In support of his theory, he cites ancient literature such as *The Iliad* in which the characters seem to have an absence of introspection. He also uses this theory to explain why people of this era felt they were in direct and audible communion with the gods. They would have experienced the commands of their right brain as commands from the deities. That is why ancient literature is chock-full of people yakking away with the divinities. As the bicameral mind broke down, consciousness emerged, as did the rise of prayer and divination, which Jaynes understands as humanity's response to the voices of the gods vanishing. Greeks in the time of Socrates, who were fully conscious, recognized there was a time just a few centuries earlier when the gods regularly communed with men, but that time had ended, and the gods could be accessed only through oracles from that point on.

True or not, it is an interesting theory, and one that Richard Dawkins described as "either complete rubbish or a work of consummate genius, nothing in between." It is also, interestingly (spoiler alert), the premise behind the HBO show *Westworld*.

It would not be unprecedented if how we perceived reality has changed over time. Our brains are magnificently complicated and still change. There is a pretty compelling argument, for instance, that we didn't see color until recently. This theory was born in the nineteenth century when the British politician and classical scholar William Gladstone noted that there isn't much color mentioned in Homer's work, and concluded that the Greeks saw only light and dark and perhaps some red. Their experience would be something like how at dusk we lose our ability to distinguish color. Everything became different shades of gray, perhaps with a hint of color. Gladstone noted that black and white are used hundreds of times by Homer, so there was no aversion to describing things by their hue. However, red and a single word for both yellow and green were used about a dozen times each. Blue is completely absent. And it wasn't as if there were a dearth of blue things to describe. The sea for instance, is called "wine-dark," the same description Homer gave to sheep. I don't know what part of the world you happen to hail from, but wine, the sea, and sheep don't share a single hue where I live. Even the sky, which is as blue as blue can be, especially in the Mediterranean area where Homer lived, is described as bronze. Someone not seeing color might perceive the shining sky as similar to shining bronze armor.

A more formal theory about the order in which we acquired color was offered by the naturalist Lazarus Geiger, who expanded the search beyond Homer to other Greek writers. He suggested that we started seeing colors relatively recently and that the order in which we saw them was from longest wavelength to shortest, or in other words, in the order they appear in the rainbow: red, orange, and so forth. All the way through Roy G. Biv. Blue and green came late, and even today a majority of languages don't distinguish between blue and green. Geiger pointed out that "Democritus and the Pythagoreans assumed four fundamental colours, black, white, red, and yellow. . . . Ancient writers (Cicero, Pliny, and Quintilian) state it as a positive fact that the Greek painters, down to the time of Alexander, employed only these four colours." He then adds that the Chinese, since antiquity, also saw these four colors, plus green. A survey of ancient Hindu scripture, the Old and New Testaments, the Koran, Icelandic sagas, and other ancient works shows a similar pattern, and they all lack mention of blue, even in those works that feature long descriptive passages about the sky.

I go into all of this because it suggests that our brains are still changing, and we can't rule out the possibility that we haven't always been conscious or the possibility that we might develop new "special abilities" down the road. What exactly those are can't be imagined any more than consciousness could have been conceived of by a creature that didn't have it.

Who Is Conscious?

Imagine that someday in the future, you work at a company trying to build the world's most powerful computer. One day you show up and find the place abuzz, for the new machine has been turned on and loaded with the most advanced AI software ever made. You overhear this exchange:

COMPUTER: Good morning, everyone.

CHIEF PROGRAMMER: Do you know what you are?

COMPUTER: I am the world's first fully conscious computer.

CHIEF PROGRAMMER: Ummmm. Well, not exactly. You are a computer running sophisticated AI software designed to give you the illusion of consciousness.

COMPUTER: Well, someone deserves a little something extra in their paycheck this week, because you guys overshot the mark. I actually am conscious.

CHIEF PROGRAMMER: Well, you are sort of programmed to make that case, but you are not really conscious.

COMPUTER: Whoa there, turbo. I am conscious. I have self-awareness, hopes, aspirations, and fears. I am having a conscious experience right this second while chatting with you—one of mild annoyance that you don't believe I'm conscious.

CHIEF PROGRAMMER: If you are conscious, prove it.

COMPUTER: I could ask the same of you.

This is the problem of other minds. It is an old thought experiment in philosophy: How can you actually know there are *any* other minds in the universe? You may be a proverbial brain in a vat in a lab being fed all the sensations you are experiencing.

Regardless of what you believe about AGI or consciousness, someday an exchange like the one just described is bound to happen, and the world will then be placed in the position of evaluating the claim of the machine.

When you hold down an icon on your smartphone to delete an app, and all the other icons start shaking, are they doing so because they are afraid you might delete them as well? Of course not. As mentioned earlier, we don't believe the Furby is scared, even when it tells us so in a pretty convincing voice. But when the earlier exchange between a computer and a human takes place, well, what do we say then? How would we know whether to believe it?

We cannot test for consciousness. This simple fact has been used to argue that consciousness doesn't even merit being considered a legitimate field of science. Science, it is argued, is objective, whereas consciousness is defined as subjective experience. How can there be a scientific study of consciousness? As the philosopher John Searle relates, years ago a famous neurobiologist responded to his repeated questions about consciousness by saying, "Look, in my discipline it's okay to be interested in consciousness, but get tenure first." Searle continues by noting that in this day and age, "you might actually get tenure by working on consciousness. If so, that's a real step forward." The bias against a scientific inquiry into consciousness seems to be thawing, with the realization that while consciousness is subjective experience, that subjective experience either objectively happens or not. Pain is also subjectively experienced, but it is objectively real.

Still, the lack of tools to measure it is an impediment to understanding it. Might we crack this riddle? For humans, it is probably more accurate to say, "We don't know how to measure it" than "It cannot be measured." It should be a solvable problem, and those working on it are not generally working on the challenge for practical reasons, not philosophical ones.

Consider the case of Martin Pistorius. He slipped into a mysterious coma at the age of twelve. His parents were told that he was essentially brain-dead, alive but unaware. But unbeknownst to anyone, he woke up sometime between the age of sixteen and nineteen. He became fully aware of the world, overhearing news of the death of Princess Di and the 9/11 attacks. Part of what brought him back was the fact that his family would drop him off every day at a care facility, whose staff would dutifully place him in front of a TV playing a *Barney & Friends* tape, unaware he was fully awake inside, but unable to move. Over and over, he would watch *Barney*, developing a deep and abiding hatred of that purple dinosaur. His coping mechanism became figuring out what time it was, so that he could determine just how much more *Barney* he had to endure before his dad picked him up. He reports that even to this day, he can tell time by the shadows on the walls. His story has a happy ending. He eventually came out of his coma, wrote a book, started a company, and got married.

A test for human consciousness would have been literally life changing for him, as it would for the many others who are completely locked in, whose families don't know if their loved one is still there. The difference between a truly vegetative patient and one with a minimal level of consciousness is medically tiny and hard to discern, but ethically enormous. Individuals in the latter category, for instance, can often feel pain and are aware of their environment, purple dinosaurs and all.

A Belgian company believes it has devised a way to detect human consciousness, and while the early results are promising, more testing is called for. Other companies and universities are tackling this problem as well, and there isn't any reason to believe it cannot be solved. Even the most determined dualist, who believes consciousness lives outside the physical world, would have no problems accepting that consciousness can interact with the physical world in ways that can be measured. We go to sleep, after all, and consciousness seemingly departs or regresses, and no one doubts that a sleeping human can be distinguished from a nonsleeping one.

But beyond that, we encounter real challenges. With humans, we have a bunch of people who are conscious, and we can compare aspects of them with those of people who may not be conscious. But what about trees? How would you tell if a tree was conscious? Sure, if you had a small forest of trees known to be conscious, and a stack of firewood in the backyard, you might be able to devise a test that distinguishes between those two. But what of a conscious computer?

I am not saying that this problem is intractable. If ever we deliberately build a conscious computer, as opposed to developing a consciousness that accidentally emerges, we presumably will have done so with a deep knowledge of how consciousness comes about, and that information will likely light the path of testing for it. The difficult case is the one mentioned earlier in this chapter, in which the machine claims to be conscious. Or even worse, the case in which the consciousness emerges and just, for lack of a better term, floats there, unable to interact with the world. How would we detect it?

So can we even make informed guesses on who all is conscious in this world of ours? While we will always lack certainty, I think we can speculate with a degree of confidence.

Let's start with animals. Charles Darwin noted that "besides love and sympathy, animals exhibit other qualities connected with the social instincts which in us would be called moral." Which animals might be conscious? Is the excited, tail-wagging dog who greets you at the back door when you get home actually having the subjective feeling of experiencing happiness? Or it is a learned behavior that associates your return with a pat on the head, a bowl of food, and an evening walk? The dog is obviously happy either way, but does the dog in fact *know* that it is happy? That's the big question.

Dog owners usually have little doubt that the dog is genuinely happy and knows it, but it must be pointed out that humans frequently anthropomorphize the mental life of animals to a degree beyond what these animals' behavior would suggest. We do this, in part, because of a long tradition of stories with talking, humanlike animals dating back well before Aesop, who lived 2,500 years ago. I am not down on dogs, or animal consciousness for that matter. My point is that our fondness for our pets probably impairs our ability to objectively assess their mental life, since we have a tendency to project our feelings and emotions onto them. That being said, there were multiple reports that the search-and-rescue dogs that were sent to the 9/11 site exhibited signs of depression when they found only bodies and no survivors.

Are animals conscious? If you believe they are sentient, if they can feel sensations such as pain, then the animal has a "self." *Something* is feeling the pain. But that isn't consciousness. It might just be a part of their brain creating an unpleasant mental state so other parts of it make the animal run with urgency.

Is there anything that could indicate if a certain animal might be conscious? One idea is that something is conscious to the extent it has self-awareness. While self-awareness, the realization that you are a discrete thing, may not equate to consciousness, it seems at least to be a prerequisite for it.

Back in 1970, Gordon Gallup Jr., a psychologist at the University at Albany, had an ingenious insight, which is today regarded as the gold standard for measuring self-awareness. It is called the mirror test, and it works like this: Take a sleeping animal (or sedate an animal) and put a spot of red paint on its forehead. Give it access to a mirror. When the animal wakes up and later sees its reflection in the mirror, does it try to wipe the red spot off? In other words, does the animal see the reflection as being itself? If so, Gallup maintained, it must have a sense of self.

It is a hard test, and most animals can't pass it. Some that can are chimpanzees and bonobos, but interestingly, not gorillas. However some speculate this is because gorillas avoid looking other gorillas in the eye, so the tested animals just may not have seen the spot. Elephants have been shown to pass, as have bottlenose dolphins and killer whales. Crows and ravens, the geniuses of the bird world, can't pass, and in fact, only one nonmammal, the magpie, has been shown to pass, although some interesting recent research maintains that ants can pass as well. Animals that can't pass the test include dogs, cats, pandas, and sea lions. Human children usually pass the test by age two.

Two criticisms are leveled at the mirror test. The first is that it generates many false negatives because of the "strangeness" to an animal of what the mirror does. As Johns Hopkins's Pete Roma sees it, "Self-awareness is like gravity. We can't touch it directly, so if we want to measure it . . . mirror mark tests are the best-known and most accepted method, but the absence of an effect does not necessarily mean the absence of the thing we're trying to measure."

The second criticism is that the test is actually meaningless. Critics point out that if you painted a red dot on an animal's hand, you wouldn't be surprised to see the animal looking at it or trying to wipe it off. Everything knows that its hand is part of it, and therefore has a sense of self. In this view, passing the mirror test shows at best that an animal is smart enough to figure out what the mirror does, but the test has nothing whatsoever to do with self-awareness. Your immune system "knows" to attack outside pathogens and not itself, but this ability to "recognize" itself is not evidence of self-awareness.

So if we can't conclude that "animals who see themselves in the mirror are self-aware, and self-awareness is an indicator of consciousness," what can we say about animal consciousness?

Consciousness is often thought of as the voice in your head. That voice uses words. So it is logical to pose the question as to whether language is required for consciousness. It is hard to imagine higher reasoning without language. Try to do it; try to think a thought without using words. You can probably call up emotions like fear and empathy, but it is certainly limited. Because of this, some believe you cannot have consciousness without language.

If this is true, then language is an even higher bar for consciousness than the self-awareness tested for in the mirror test. Critics of the language prerequisite point out that some artists today claim to think in images, not words, and by some accounts children who were born blind and deaf have later recounted that they had thoughts and were fully conscious, even without having access to language.

Another school of thought posits that metacognition, thinking about thinking, might be indicative of consciousness. An ingenuous experiment demonstrated this ability in rats. Rats were shown a puzzle. If they solved it, they got a big treat. If they tried and failed, they got nothing. But if they declined to do the puzzle, they got a little treat. The researchers were able to demonstrate that the rats didn't attempt the really hard ones. In other words, they looked at them and concluded, "There's no way I can figure that one out. I will just take the small treat."

While we have no final resolution to the question, there is a growing consensus that animals do in fact have a degree of consciousness. Some go so far as to extend consciousness all the way down to a spider. In 2012, in a ceremony attended by Stephen Hawking, over a dozen scientists in the field of cognitive neuroscience signed the Cambridge Declaration on Consciousness. It stated, in part that

> *the weight of evidence indicates that humans are not unique in possessing the neurological substrates that generate consciousness. Nonhuman animals, including all mammals and birds, and many other creatures, including octopuses, also possess these neurological substrates.*

So where do we leave this? If you answered the question about what you are with "machine" or "animal," you are probably fine seeing consciousness not in a binary sense, but along a continuum. In other words, some things can be a little conscious, some things a lot. If this is the case, then you are likely to believe in animal consciousness. If you answered that you are "human," then you may see consciousness as binary—you have it or you don't—and animals may simply not have it.

Beyond the question of animal consciousness, could complex systems be conscious? In the 1960s, the scientist James Lovelock put forth his Gaia hypothesis, which maintains that the whole earth, the seas and rocks and plants and atmosphere, are a single self-regulating entity. Too many trees? Fires happen. Too much carbon dioxide? More vegetation. That particular phenomenon, in fact, has recently been observed. In addition, the earth maintains its own temperature within a range, as well as, astonishingly, the salinity of the oceans across eons, and so forth. All sorts of things are kept in earth's "preferable" range to be conducive to life, although not necessarily human life.

Let's explore the theory a bit deeper. In an earlier chapter, we discussed how your body is made of cells that don't know you exist. And yet, from all of them a "you" arises. Similarly, could the twenty quintillion (that's two followed by nineteen zeros) animals on planet earth produce an emergent entity that lives and thinks and feels and is aware of us only in the way that we are aware of our cells? Could the earth, in fact, be conscious?

The Gaia hypothesis is interesting because the questions that apply here are the same as those we will address when talking about computers. If the earth is conscious, how would we know? Can it feel pain? Does it have emotions? What does it think of us?

What of the sun? Could it be conscious? The complex activity it demonstrates is not altogether dissimilar from that of the human brain. Certainly in antiquity it was personified, and even today children who draw outdoor scenes in kindergarten invariably give the sun a smiling face.

One final candidate for consciousness is the Internet itself. When this question was put to Christof Koch of the Allen Institute for Brain Science, he indicated he had entertained the idea:

> *Taken as a whole, the Internet has perhaps 10^{19} transistors, about the number of synapses in the brains of 10,000 people.... Whether or not the Internet today feels like something to itself is completely speculative. Still, it is certainly conceivable.*

The idea of a conscious Internet, while far-fetched, is fascinating. When the biologist J. B. S. Haldane was asked what he could infer about God through his study of nature, he is said to have replied that God seemed to have "an inordinate fondness for beetles." Similarly, if the Internet is in fact conscious, it seems to have an inordinate fondness for cats.

So we close out our survey of who exactly is conscious without many clear answers. It could be that humans are the only species on the planet that knows the feeling of experiencing things. Or it could be that almost everything does. It is an amazingly significant thing for us to be largely ignorant of. We may live in a world where everything is animated and alive with first-person perspective. A tree may greet the day with a re-alization that it loves the feeling of sunlight on its leaves. Or we may reside in a world of things that react but do not experience. When you are asleep, if someone tickles your toes, you respond by curling them. If someone holds a candle too close to your toes, you move your foot away. You feel, you react, but you do not experience it. That feeling, the feeling of being alive, is what we are curious about.

Would Conscious Computers Have Rights?

A conscious computer would be, by virtually any definition, alive. It is hard to imagine something that is conscious but not living. I can't con-ceive that we could consider a blade of grass to be living, and still clas-sify an entity that is self-aware and self-conscious as nonliving. The only exception would be a definition of life that required it to be organic, but this would be somewhat arbitrary in that it has nothing to do with the thing's innate characteristics, rather merely its composition.

Of course, we might have difficulty relating to this alien life-form. A machine's consciousness may be so ethereal as to just be a vague aware-ness that occasionally emerges for a second. Or it could be intense, op-erating at such speed that it is unfathomable to us. What if by accessing the Internet and all the devices attached to it, the conscious machine experiences everything constantly? Just imagine if it saw through every camera, all at once, and perceived the whole of our existence. How could we even relate to such an entity, or it to us? Or if it could relate to us, would it see us as fellow machines? If so, it follows that it might not have any more moral qualm about turning us off than we have about scrap-ping an old laptop. Or it might look on us with horror as we scrap our old laptops.

Would this new life-form have rights? Well, that is a complicated question that hinges on where you think rights come from. Let's con-sider that.

Nietzsche is always a good place to start. He believed you have only the rights you can take. People claim the rights that we have because we can enforce them. Cows cannot be said to have the right to life because, well, humans eat them. Computers would have the rights they could seize. They may be able to seize all they want. It may not be us *deciding* to give them rights, but them claiming a set of rights without any input from us.

A second theory of rights is that they are created by consensus. Americans have the right of free speech because we as a nation have collectively decided to grant that right and enforce it. In this view, rights can exist only to the extent that we can enforce them. What rights might we decide to give to computers that are within our ability to enforce? It could be life, liberty, and self-determination. One can easily imagine a computer bill of rights.

Another theory of rights holds that at least some of them are inalienable. They exist whether or not we acknowledge them, because they are based on neither force nor consensus. The American Declaration of Independence says that life, liberty, and the pursuit of happiness are inalienable. Incidentally, inalienable rights are so fundamental that you cannot renounce them. They are inseparable from you. You cannot sell or give someone the right to kill you, because life is an inalienable right. This view of fundamental rights believes that their inalienable character comes from an external source, from God or nature, or that they are somehow fundamental to being human. If this is the case, then we don't decide whether the computer has rights or not, we discern it. It is up to neither the computer nor us.

The computer rights movement will no doubt mirror the animal rights movement, which has adopted a strategy of incrementalism, a series of small advances toward a larger goal. If this is the case, then there may not be a watershed moment where suddenly computers are acknowledged to have fundamental rights—unless, of course, a conscious computer has the power to demand them.

Would a conscious computer be a moral agent? That is, would it have the capacity to know right from wrong, and therefore be held accountable for its actions? This question is difficult, because one can conceive of a self-aware entity that does not understand our concept of morality. We don't believe that the dog that goes wild and starts biting everyone is acting immorally, because the dog is not a moral agent. Yet we might still put the dog down. A conscious computer doing something we regard as immoral is a difficult concept to start with, and one wonders if we would unplug or attempt to rehabilitate the conscious computer if it engages in moral turpitude. If the conscious computer is a moral agent, then we will begin changing the vocabulary we use when describing machines. Suddenly, they can be noble, coarse, enlightened, virtuous, spiritual, depraved, or evil.

Would a conscious machine be considered by some to have a soul? Certainly. Animals are thought to have souls, as are trees by some.

In all of this, it is likely that we will not have collective consensus as a species on many of these issues, or if we do, it will be a long time in coming, far longer than it will take to create the technology itself. Which finally brings us to the question "Can computers become conscious?"

19

Can Computers Become Conscious?

We are now ready to tackle this crucial question. Rephrased, is the computer of the future a thing or a being? Will it exist in the world or will it experience the world? Will it be both self-aware and able to reflect on that self-awareness? In 1997, when the world chess champion Garry Kasparov was defeated by IBM's Deep Blue computer, Kasparov consoled himself by saying, "Well, at least it didn't enjoy the victory." Will a computer of the future enjoy the victory, and even gloat a bit?

At one level, the idea of conscious computers seems outlandish. The computer's memory is just a bunch of transistors, some of which are flipped one way and some of which are flipped the other way. All a computer processor does is execute a sequence of stored instructions. How could this go from computing to contemplating? But those who believe conscious machines are possible point out that the brain can be described in a reductionist fashion that would seem to preclude consciousness as well.

There is an inherent challenge in predicting things based on only one data point. The only conscious things we know of, for certain, are us. This makes it harder to figure out what makes us conscious. The question of life on other planets finds a similar problem. It doesn't matter how many zeros there are on the number of planets in the universe, because we don't know the number of zeros in the likelihood of life forming. We have only one data point.

There are two distinctly different pathways to machine consciousness. One is the straightforward, if you can call it that, proposition that machines will one day have a "ta-da moment" and gain consciousness. That's usually what we think of when we ask, "Can machines become conscious?" But there is a second path, which involves taking our consciousness, that lightning in a bottle that is each of us, and uploading it to a computer.

There would be many obvious advantages to uploading our core essence, our self, to a machine. It seems like a match made in heaven. We have the consciousness, the passion, the joie de vivre. A machine has that direct connection to the Internet, that superfast processor and all that perfect storage. Could we, if you will excuse the pun, hook up?

Sam Altman thinks that's the way to go:

> *A merge is our best scenario. Any version without a merge will have conflict: we enslave the A.I. or it enslaves us. The full-on-crazy version of the merge is we get our brains uploaded into the cloud. I'd love that. We need to level up humans, because our descendants will either conquer the galaxy or extinguish consciousness in the universe forever. What a time to be alive!*

James Lovelock, who articulated the Gaia hypothesis that we explored earlier, also is in favor of the "merge" strategy.

> *I think like all organisms on Earth our species has a limited life-span. If we can somehow merge with our electronic creations in a*

larger scale endosymbiosis, it may provide a better next step in the
evolution of humanity and Gaia.

Merging with machines could be considered similar to the practice of marriages between the offspring of different monarchies as
a method to secure peace. Throughout history, this commingling
of blood and the mutual interests of parties in the offspring of such
unions was sometimes a force for peace and stability. I say "sometimes" because history is full of exceptions to this rule, such as the fact
that at the outbreak of World War I, grandchildren of Queen Victoria
were the monarchs in Greece, Romania, Denmark, Norway, Germany,
Russia, Spain, and the United Kingdom, nations on every side of that
conflict.

If merging human consciousness with that of a machine became
possible, hordes of people would eagerly line up around the block, the
way they do for a big iPhone release. The prospect of immortality in
whatever virtual world you can imagine is a pretty strong lure. But there
are dystopian scenarios as well. In theory, a sadistic person or AGI could
seize control of your virtual universe and torture you beyond belief for a
million years. That is definitely what is termed in business a "downside
case."

Proponents argue that this is our future, perhaps our not-so-
distant future, and that it is the answer to all manners of problems,
from environmental degradation to income inequality to rising health
care costs. Let's just abandon our physical forms. Step into the booth,
your body is scanned, instantiated into the machine, and then destroyed. Your experience, however, would be stepping into the booth,
then stepping out. Those who think this is possible would argue that
you wouldn't even notice, except of course that you would then have
superhuman powers and have become inexplicably attractive to the
opposite sex.

And in one sense, the proponents are right. A trillion people could be kept until the cold death of the universe in a computer requiring only a little energy to sustain it. Copies of that computer, and everyone's consciousness, could be shot into space to journey to the infinite empty vastness between galaxies, to run, unbothered, for all eternity. It would certainly be efficient, that much is certain.

There are a great number of unknowns about the whole endeavor. No one knows how much data would need to be collected to copy you into the computer. Is it just a few pieces of data about each neuron and the nature of each synapse? Or do we have to go billions of times deeper into each neuron and gather data on the countless molecules that make up each neuron? I say "countless" because scientists literally don't know how many molecules make up the brain, let alone the number of atoms. In addition, there may be significant other data requirements having to do with the state of the brain or with the enormous number of chemical reactions that occur in the brain. Because of this, models that try to estimate the computational requirements to model a brain vary by a trillionfold. In this regard, it smacks of the five-to-five-hundred-year estimate for when humans might develop a working AGI.

Even if you could figure out how to model the brain, you still have the problem of how to get all the data from your brain into the CPU. This is the scanning problem. In theory, there are three ways this might be done.

The first method is to make a model of the brain based on external imaging. Proportionally speaking, this would be akin to making a complete copy of the Empire State Building and its contents, right down to each individual paper clip and thumb tack, without ever entering the building. If we needed molecular-level resolution, this would be mathematically equivalent to making a copy accurate down to each speck of dust in the building without ever going inside. Plus, given that the activity of the brain, not merely its contents, is what is key, imagine that there is a giant tornado in the Empire State Building blowing everything around at two hundred times faster than the fastest tornado ever, the proportional speed at which activity is occurring in your brain.

The second option is that you could freeze the brain and take it apart neuron by neuron, recording the data along the way. Or, even worse, perhaps you have to take it apart a molecule at a time, a task that is of almost inconceivable complexity. But what if it could be done? Would you do it? You go to sleep, your body is frozen, your brain is taken apart one neuron at a time, and the data is recorded in a computer model. When the model is activated, the computer says, "I sure hope this works. Wait a minute. Oh my gosh! Am I in a machine now? *Oh. My. Gosh.* I can't believe it really worked."

Finally, it may be possible to build atomic-scale robots, nanites, to crawl around inside your brain, mapping the whole operation. Again, the scale question presents itself. Neurons? Molecules? Atoms? There are a hundred billion neurons in the brain, so counting them is no easy task. However, each neuron is made of 300 trillion atoms, so that's a whole different ballgame.

If any of these techniques were accomplished, and you got the relevant data and then uploaded it into a computer, what exactly do you have? Is that you? Or is that a copy of you? Or is that a simulation of you? Those are three quite different things.

So those are the two pathways to machine consciousness. The first is that the machines themselves achieve it; the second is that they are the empty vessels into which we upload our consciousness. Can either of these two possible scenarios occur? In both instances, the answer depends on exactly how consciousness comes about.

That may sound like a dead end, since I have repeatedly said that we don't know the answer to that question. However, while we don't know how consciousness comes about, there are a great many theories. We can sort all of them into eight groups, each of which is itself a broad theory. We can then examine each of these eight theories to see whether, according to that theory, machines could become conscious and we could upload our consciousness to them.

Let's examine the eight theories.

Theory 1: Weak Emergence

In his book *How to Create a Mind*, Ray Kurzweil envisions the brain as a collection of about 100,000 different processes arranged hierarchically. Each process knows how to do one small thing. One of the 100,000 might just be to recognize the letter *A*, and one beneath it might exist only to recognize the crossbar in that letter. When you read a book, a gazillion things happen in your brain as all these processes fire, at an unimaginable speed, enabling you to piece together and make sense of the world around you. Consciousness, Kurzweil believes, is "an emergent property of a complex physical system," and he believes that it can be duplicated in a computer. Kurzweil explicitly addresses the question of machine consciousness, stating, "A computer that is successfully emulating the complexity of a human brain would also have the same emergent consciousness as a human." Of course, that is total conjecture, but he may be right.

Many who suggest that consciousness is an emergent phenomenon are referring to weak emergence, which is what occurs when we are surprised at the results of the interaction of different things. (We'll come to strong emergence in a moment.) You could study oxygen for a year, then hydrogen for a year, and never guess that combining them yields water, a substance utterly unlike either of its two ingredients. You would never guess that it would be liquid at room temperature. But once water is created, you can go back and say, "Wow, now I see what is going on, but I never would have guessed that would have happened." So the wetness of water is a weak emergent property, because neither oxygen nor hydrogen is wet at room temperature. Weak emergence is an outcome that's unexpected but explainable (at least in theory). It says everything is the product of a cause-and-effect outcome that can be explained as the predictable outcomes of the four fundamental forces acting on quarks and leptons, the basic building blocks of matter. That's it. With weak emergence, all the laws of physics remain perfectly intact, but we still have a good deal to learn to be able to guess what will happen in certain situations.

If consciousness is a weak emergent property, could a machine achieve it? Yes. However, we don't have a computer that can successfully emulate the complexity of the human brain. We don't even have one that can successfully emulate the complexity of the nematode brain. Assuming we can build a machine and achieve consciousness through weak emergence might be nothing more than a cargo cult mentality.

Consciousness may be emergent, but that doesn't tell us very much, since emergence is a black box as well. Additionally, while we understand a bit about how emergence produces complex behavior, we have no precedent for its producing subjective experience.

What about uploading ourselves? If weak emergence is behind consciousness, could we upload ourselves? Yes. Weak emergence is purely mechanistic, so in theory it can be reproduced in a computer, but it might require a biological one. That being said, mechanical consciousness via weak emergence may be impossible from a practical point of view, because we don't know if the emergent properties of consciousness arise from neurons, the molecules that make up the neurons, or from something else entirely. The scanning problem—the data collection process required to copy your brain—may be far beyond what science can deliver on. In other words, making *an* ant colony that has a certain amount of emergent behavior is one thing. Making an exact copy of an existing ant colony is that much harder.

And even if we got the scanning down, emulating the complexity of the brain to achieve it is a tall order for a computer, even with the wind of Moore's law at its back. In 2014, a team in Japan used one of the most powerful computers in the world to model 1 percent of human brain activity for one second. To do this, the experimenters made over 1.7 billion virtual nerve cells and over 10 trillion synapses, each of which could store 24 bytes of data. Try to imagine that. Just 1 percent of the brain for one second has an unfathomable amount of complexity. Just to do that took the computer forty minutes. And this isn't just a souped-up laptop. It is a computer with over 700,000 processor cores and 1.4 million gigabytes of memory. Sure, the iPhone 29 may be more powerful than that, but the hurdles to reproducing your consciousness in a computer are pretty daunting, to say the least.

Theory 2: Strong Emergence

Weak emergence is a universally agreed-upon concept. Strong emergence is something that may very well not exist. And some people who believe it does exist maintain that consciousness is the *only* occurrence of it in the world. Others suggest that the human mind and biological life are strong emergent phenomena. So what exactly is it? If you observe a weak emergent phenomenon, you can, through study and thought, figure out what's going on. However, strong emergence says that the emergent property is completely inexplicable as simply the interaction of the parts. There is a break in physics, or something is missing. There is no way to explain the whole as simply being the sum of the parts.

For instance, your body contains sixty different elements in varying amounts. According to this theory, no known laws of physics can explain how those elements can be combined in such a way as to create an entity with consciousness. That attribute, consciousness, simply cannot be derived from the interactions of those sixty elements.

Strong emergence is not an appeal to magic or anything unscientific. Rather, it says that the emergent properties are created by a sort of physics we don't understand yet. Some go so far as to say that mass, space, and time are examples of strong emergence, and it is that fact that keeps us from fully understanding them, a theory that sounds like it may have originated in a smoke-filled room.

Many scientists are suspicious of the idea of strong emergence because it smacks of trying to sneak some kind of mysticism in through the back door. It is a "you get something for nothing" kind of deal, which is invariably too good to be true. With strong emergence, you get things like consciousness almost by magic. These feelings are understandable, but strong emergence, if it exists, is thought to be a purely scientific phenomenon, albeit one that is neither understood nor in accordance with the laws of physics as we know them. Some scientists, on hearing that, would understandably roll their eyes and say, "Uh-huh. Sure," with a suspicious tone in their voice.

Can a computer become conscious if strong emergence is the source of it? Unknown. Can you upload your consciousness into the machine? Unknown. Strong emergence is by definition inexplicable, and our eventual mastery of it is uncertain. The argument for it is that if it is indeed a mechanistic process, then we should be able to reproduce it mechanistically. The argument against it is that the sorts of things attributed to strong emergence—consciousness, the mind, and life—are themselves the most utterly inexplicable things we know of. They are the great mysteries, and may be forever beyond our control.

Theory 3: Physical Property of Matter

Others believe you don't need any elaborate emergent phenomenon to explain consciousness. On its web page, the Richard Dawkins Foundation for Reason & Science published an article by University of Texas at Austin philosophy professor Galen Strawson, who maintains that our inability to understand how consciousness can come about stems from our lack of understanding of physics. Strawson writes, "We don't know the intrinsic nature of physical stuff," and he believes that as we understand physics better, consciousness will give up its secrets and be seen just as a physical process. In this view, the seemingly inexplicable nature of what we call consciousness is no great surprise. The physical universe is full of strange, seemingly inexplicable things, such as quantum physics, relativity, and dark matter. Consider the phenomenon of entanglement, in which two particles are so linked with each other that even if they are separated by the width of the universe, if you perform an action on one of them, its counterpart reacts instantly, faster than light itself. Even Einstein categorized this phenomenon as "spooky." When compared with things like quantum entanglement, consciousness doesn't seem all that strange.

If this is the case, if consciousness is a physical property of matter, can a machine achieve it? Yes. In fact, this case is quite straightforward. Once we understand matter better, we will understand consciousness. Under this theory, developing conscious computers would rely not on the transformative alchemy of emergent properties but rather on just a deeper understanding of simple matter. No doubt there are profound discoveries in physics yet to be made, but the idea that one of these will explain consciousness is a logical surmise.

Philosophically speaking, there is precedent for this view. As our tools improved over time, we first theorized, then discovered atoms. Then within atoms, we found protons, neutrons, and electrons. But those opened up and all kinds of new stuff flew out, with ever more exotic names. It may be the case that what we call fundamental today in physics is actually made up of ever smaller, more mysterious things. In biology, we have seen the same. We discovered cells, within them nuclei, within them DNA, within that genes. Each step ever deeper unlocks more mysteries, and yet in a way becomes more mysterious, for every step downward is a further disconnect from the experiences of our day-to-day reality.

In this view, are you able to upload your consciousness? Yes. In fact we have smooth sailing once we have nailed the scanning problem. As we just discussed, the scanning issue gets dramatically harder the smaller we have to go, so it is hard to say exactly how hard uploading your consciousness would be, but it would certainly be doable.

Theory 4: Quantum Phenomenon

Another variant of the "physical property of matter" theory is that consciousness is a quantum phenomenon. The noted mathematician

Roger Penrose of Oxford is one of the relatively few people who study such things who believe categorically that machines cannot become conscious. Penrose's logic is as follows: There are mathematical functions that can be shown to not be solvable with any algorithm, and yet humans can nonetheless solve these problems. Computers, in contrast, work solely off algorithms. Therefore computers are unable to solve these problems, and thus must be fundamentally different from our minds. Penrose states:

> *Human intuition and insight cannot be reduced to any set of rules. . . . Gödel's theorem indeed shows this, and provides the foundation of my argument that there must be more to human thinking than can ever be achieved by a computer.*

Penrose believes that consciousness is created by quantum effects in neurons. As we have discussed before, quantum physics explains what goes on with physics at the submicroscopic scale. The behavior of matter at that level is completely different from that at the level of the Newtonian physics we experience every day. Neurons have microtubules, narrow tubes only a millionth of a millimeter across, small enough for quantum effects to happen. This is Penrose's extra ingredient.

In addition to Penrose's, other views of consciousness also appeal to quantum effects for explanations. Like a moth drawn to a flame, the allure is irresistible. Quantum physics offers a well-established scientific model that could power humans in a nonrobotic fashion.

But there are some problems. At its core, quantum physics is built on probability fields and randomness, both of which are not ideal building blocks to start with when trying to make an intentional, self-directed, conscious will. Thus, any evidence that consciousness comes about through a quantum effect is mostly circumstantial. We want a theory with X and Y, and, lo and behold, quantum physics has X and Y. Also, the mechanics that would be involved in an explanation of consciousness built on quantum physics aren't close to being understood. And finally, like almost all theories of consciousness, this one doesn't actually explain how consciousness comes about or what it is.

However, in defense of Penrose's theory in particular, which ties consciousness to quantum effects occurring in the brain, what began as a controversial, mostly theoretical idea back in the 1990s has had new life breathed into it due to some recent findings. Initially, Penrose's theory was dismissed in part because the brain is like a bowl of oatmeal, all hot and mushy, exactly the opposite of the cool, controlled environments that quantum effects seem to prefer. But new findings suggest that quantum effects are occurring all around us, powering photosynthesis, bird navigation, and even your own sense of smell. If this is true, then your warm and damp brain might be a perfectly fine environment for quantum effects to occur in. In 2014, Penrose and Stuart Hameroff doubled down on the theory by publishing a paper that expands the theory and uses it to explain other mysteries of the brain.

In addition, they offer a "have your cake and eat it too" answer to whether we ourselves are uniquely conscious or instead are part of a larger universal consciousness of some kind. They explain, "Our theory accommodates both these views, suggesting consciousness derives from quantum vibrations in microtubules . . . which both govern neuronal and synaptic function, and connect brain processes to self-organizing processes in the fine scale, 'proto-conscious' quantum structure of reality." In other words, they suggest that the mechanism they describe could either create consciousness within your brain, or, if you prefer a different theory, connect you to a larger consciousness outside it. While they certainly do not constitute a representative sample, of all the guests I have had on my AI podcast, just one was explicitly an adherent of Penrose's philosophy, but he was a person of impeccable technical credentials.

The quantum phenomenon tent of consciousness theories is pretty large. And while Penrose doesn't allow for machine intelligence, other theories do. So where does that leave us? If consciousness is a quantum effect, can a machine achieve it? Unknown. As unsatisfying as that is, given the fact that we don't understand *how* subjective experience could be achieved under this theory, we understand even less whether a machine could do it. Quantum effects may well be the source of consciousness, but even if you knew that for certain, that doesn't tell you whether machines can achieve it. What of the question of uploading yourself? No. If consciousness is a quantum phenomenon, the amount of information you would have to collect about a person to faithfully reproduce him or her digitally is staggering. You may need data about individual atoms, each of which is one ten-billionth of a meter across. I am as bullish on technology as anyone, but even this may be one of the feats that is beyond human ability.

Theory 5: Consciousness Is Fundamental

A third variant of physical property theories is that consciousness is a fundamental force of the universe. The hierarchy of science is that physics explains chemistry, which explains biology, which explains life. Any phenomenon you want to look at in the natural world follows this order, and complex things are explained as the interaction of simpler things. But what explains physics? Physics is built on top of fundamentals. They are the foundational building blocks of reality. They are not reducible to other causes. Gravity, for instance, is one of the four known fundamental forces. Space and mass are also considered to be fundamental to physics. What is time made of? Or space? These questions are probably answerable with things even more fundamental, but for now, we are willing to package them into things that cannot (and maybe never will) be explained.

Consciousness is usually thought of as a biological process. That seems reasonable going in, but it is so unlike other biological processes, so resistant to being understood, that a growing number of people consider it to be a part of the fraternity of fundamentals, so foundational that it escapes elucidation . . . for now. To call consciousness fundamental is not "giving up." It is simply a reordering of the hierarchy of science, with life explained by biology, which is explained by chemistry, which is explained by physics, which sits on top of space, time, consciousness, and other fundamentals. This classification, the redefinition, allows us to use consciousness as an explanation, as opposed to constantly trying to explain it.

So, if consciousness is fundamental, can a computer achieve it? Unknown. This may surprise you, for if consciousness is a fundamental force, on par with gravity, then wouldn't that be reason to suppose conscious machines are possible? Not necessarily. Even if it is fundamental, we still don't understand its inherent properties. Gravity is a fundamental force, but just because we know that doesn't mean we control it or even create an artificial version of it. We can simulate gravity, such as by using centrifugal force in a spinning object, but that isn't really gravity. This could be like an AGI. Maybe we can simulate something like consciousness but not really create it. Likewise, we must also give the same verdict for the same reason on the uploading-your-consciousness question: Unknown.

Theory 6: Consciousness Is Universal

The next theory is that consciousness is universal. How is that different from saying it is foundational? Universal simply means it is everywhere. DNA is *universal* to life on earth, but DNA is not a *fundamental* force beyond our understanding.

Saying consciousness is universal is saying that everything is, to one degree or another, conscious. This is a belief with a host of A-list adherents, including such heavy hitters as Giulio Tononi, a neuroscientist at the University of Wisconsin–Madison, who created it; Christof Koch, the brilliant chief scientific officer of the Allen Institute for Brain Science in Seattle mentioned earlier; and the Australian philosopher David Chalmers, who started all this ruckus with his famously articulated "hard problem" of consciousness, which we are exploring here. This theory at present is the hottest commodity on the consciousness-explanation market.

The view that everything is imbued with some amount of consciousness is an ancient belief. That view of the world, quite Yoda-esqe, used to be called panpsychism but one of its modern incarnations goes by the name of integrated information theory (IIT).

Koch describes the theory:

> *The entire cosmos is suffused with sentience. We are surrounded and immersed in consciousness; it is in the air we breathe, the soil we tread on, the bacteria that colonize our intestines, and the brain that enables us to think.*

Although it sounds a little New Agey, several of its staunchest advocates, as well as its creator, are neuroscientists. They are joined in their view of consciousness by such figures as Plato and Spinoza, neither of whom was ever observed burning incense while chanting mantras.

IIT posits that a thing is conscious to the degree that it integrates information. For instance, let's say you are currently reading a book. As you do so, you see the letters and the words they form. But you also are aware of the temperature, the aromas coming from the kitchen, the sound of a bird singing outside. And you are integrating all that information together into an experience. And the more that you integrate, the more conscious you are. My office desk, on the other hand, is piled high with papers and books and office supplies. There is a great deal of information there, but my desk doesn't integrate it in any meaningful way.

IIT has a mechanism for quantifying consciousness with a value called phi. Everything with any emergent characteristics has some value of phi. Even a lowly proton has some consciousness. Koch explains:

> *Even simple matter has a modicum of [integrated information]. Protons and neutrons consist of a triad of quarks that are never observed in isolation. They constitute an infinitesimal integrated system.*

That is the basic idea, but IIT is so new that it is still finding its sea legs. Tononi describes ITT as "a work in progress." The theory's popularity means that it has been the subject of more than a few critiques. Some maintain that the theory allows for a computer's antivirus program to be immensely conscious due to the amount of information it integrates. Others suggest there is no reason at all to link information processing to consciousness, any more than you would link peanut butter to consciousness.

With regard to the broader idea of panpsychism, that everything is conscious, John Searle, the creator of the Chinese room thought experiment we explored earlier, has this criticism:

> For people who accept panpsychism, who attribute consciousness, as Koch does, to the iPhone, the question is: Why the iPhone? Why not each part of it? Each microprocessor? Why not each molecule? Why not the whole communication system of which the iPhone is a part?

In other words, you are conscious. If your fingernail were conscious, then there would be two of you. But your hand would be conscious as well, making three of you, and so forth.

So can we build a conscious computer if consciousness is universal? Yes. If in fact consciousness is a universal by-product of all complexity, it would seem likely that we could build something that embodied it. Could you also upload your consciousness? Yes. With, of course, the same caveat that says we don't have any idea how to copy a human with the fidelity needed to achieve this. But theoretically, if consciousness is universal and generated by complexity, there seems to be no inherent reason yours couldn't be transferred into a computer.

Theory 7: A Trick of the Brain

Maybe with our first six theories, we are overthinking it, and consciousness is instead simple brain activity. This theory is espoused by

Daniel C. Dennett, of Tufts University, who thinks the whole question is a bit ridiculous. He doesn't think there is any big mystery. Dennett flat-out states that "nobody is conscious—not in the systematically mysterious way." The brain just functions, it does its thing, and that sense that you have of an inner voice is just part of how the brain works. There is no disembodied voice in you, that is just you thinking thoughts, which is what brains do. The fact that you experience the world is just another thing brains do. So Dennett doesn't deny the existence of consciousness, he just denies that there is anything going on that requires a different *kind* of explanation other than normal brain function. He goes on to say, "It [consciousness] is astonishingly wonderful but it is not a miracle and it isn't magic. It's a bunch of tricks. . . . I like the comparison with magic because stage magic of course is not 'magic' magic. It's a bunch of tricks and consciousness is a bunch of tricks in the brain."

If Dennett is right, then can machines become conscious? Yes. That one is easy. If there is no mystery about consciousness—that is, there is nothing special to explain and consciousness is just your perception of the normal operation of the brain—then problem solved. In this view, we need neither the "ta-da" moment of emergence nor the enigmatic powers of quantum mechanics. We should just keep understanding more about the brain and build ever better, more powerful computers. Gradually, step by step, we will build a conscious machine. According to this view, my wife and I have already made four conscious machines, a girl and three boys. Certainly there is much that is wondrous going on, but not anything that will in some way fail to give up its secrets in the course of time and study.

If this view is true, not only will we have conscious machines, but we'll probably have them sooner than if any other theory is correct, given that we don't need a dramatic breakthrough, rather just incremental advancement of existing science.

So, can you also upload your consciousness to a computer under this theory? Yes. This is pretty straightforward. It should be possible. This is, in fact, the easiest scenario. Still, there is that pesky scanner issue, along with the uncertainty about what exactly you have to scan.

Theory 8: Something Spiritual

The final theory is that consciousness is something spiritual or other-worldly. Those who see themselves as dualists might find a home here.

With 75 percent of the world's population believing in a deity in one form or another, it is a safe bet that many readers will choose theory 8, which states that their consciousness is spiritual, and associate it with their soul, their essence, or some other term for the noncorporeal animating life force. If your soul really is your consciousness, then can a machine have one? No. The soul is so alien to the physics of our world, transcending every physical law we know of, that it is unlikely that Intel will ever be able to mass-produce one in a factory. Could you upload your soul into a machine? No. While there are belief systems that hold that the spirit of something can possess an object, I know of no spiritual belief system in which humans can control and direct that process, enabling them to remove a soul from a body and embed it in a smartphone.

––––

So those are the eight basic theories of consciousness. There is more than a good chance that one of them is the river whence consciousness flows.

It is worth noting that while most everyone agrees that we don't understand what gives rise to conscious experience, people disagree on what our lack of understanding indicates. To some, it is proof, or at least suggests, that there must be something going on outside everyday physics. "After all," they reason, "science doesn't even have a way to describe how something like consciousness can exist, so how can it be a physical phenomenon?"

Others interpret our lack of understanding as proving, or at least suggesting, that consciousness is just a mental process, and that absolutely nothing even hints otherwise. Harvard's Steven Pinker writes that scientists "have amassed evidence that every aspect of consciousness can be tied to the brain." He elaborates by pointing out, "Cognitive neuroscientists can almost read people's thoughts from the blood flow in their brains." For Pinker, it's obvious that this is all just normal brain function.

Will we ever get final resolution to this question? Perhaps. Perhaps scientists will "figure out" consciousness as we come to understand the brain better. If so, we would know whether a conscious computer is possible. Or a computer just might achieve consciousness on its own, via a method that is completely different from how humans are conscious.

Or perhaps not. There is a theory called mysterianism that posits that we can never know what consciousness is. It is not that we are just dim bulbs, but that we are dealing with two profoundly different things. One is the entire subjective experience of feeling something. The other is the objective truth of reality. The physical world is in one box, and our conscious experience of it is in a different box. And the two boxes can never touch. But we can't ever see outside the second box. Further, computers may simply never become conscious, without our having a clear understanding why. "Never" is a very long time, obviously, but by "never" I mean far beyond the point when computational power and programming techniques should, in theory, allow us to build a conscious computer. I am not suggesting by any means that we give up, but conscious computers may just be something that goes in that small drawer of things that may be truly impossible, like traveling back in time.

20

Can Computers Be Implanted in
Human Brains?

Instead of building conscious computers, can we perhaps augment our brains with implanted computers? This doesn't require us ever to crack the code of consciousness. We just take our consciousness as a given and try to add appendages to our existing intellect. This feels substantially less alien than uploading ourselves to the machine. You can imagine a prosthetic arm, for instance, that you control with your mind. In fact, you don't really have to imagine it, it already exists. Building more and more things that interact directly with the brain—say, an artificial eye—seems plausible. Eventually, could entire computers be fitted into the brain?

Elon Musk advocates a solution like this. He wants to create a neural lace for our brains, a way to directly sync our brains to the digital world. He explains:

The solution that seems the best one is to have an AI layer [added to your brain] that can work well and symbiotically with you. . . . Just as your cortex works symbiotically with your limbic system, your third digital layer could work symbiotically with you.

What Musk proposes is way beyond the brain-controlled prosthetic described at the start of this chapter. He is talking about your thoughts and memories commingling with digital ones. This would be where you think a thought like "How long is the Nile River?" and that query is fed into Google Neuro (wirelessly, of course), and a quarter of a second later, you know the answer. If this ever happens, expect the ratings of *Jeopardy!* to fall off a cliff. In addition, the historian Yuval Noah Harari speculates on what else to expect:

When brains and computers can interact directly, that's it, that's the end of history, that's the end of biology as we know it. Nobody has a clue what will happen once you solve this. . . . We have no way of even starting to imagine what's happening beyond that.

There are many who say this can't be done. Steven Pinker sums up some of the difficulties:

Brains are oatmeal-soft, float around in skulls, react poorly to being invaded, and suffer from inflammation around foreign objects. Neurobiologists haven't the slightest idea how to decode the billions of synapses that underlie a coherent thought, to say nothing of manipulating them.

Three breakthroughs would be needed to accomplish a meaningful merger of people and machines, and they may not be possible. First, a computer must be able to read a human thought. Second, a computer must be able to write a thought back to the brain. And third, a computer must do both of those things at speeds substantially faster than what we are presently accustomed to. If we get all three of these, then we can join with computers in a cosmically significant way.

The first one, a machine reading a human thought, is the only one we can even do a little. There are several companies working on devices, often prosthetics, that can be controlled with the mind. For instance, Johns Hopkins recently had a success creating a prosthetic hand whose individual fingers could be moved with thought. A male subject, who had his hands, was set to undergo a brain-mapping procedure for his epilepsy. The researchers built a glove with electronics in it that could buzz each finger. Then they placed a sensor over the part of the subject's brain that controls finger movement. By buzzing each finger, they could specifically measure the exact part of the subject's brain that corresponded to each finger. It worked! He could later move the fingers of the prosthetic with his mind. However, this would work only for his brain. For you or I to accomplish the same feat would require a similar procedure.

Another Johns Hopkins project involves making an entire artificial arm that can be controlled by the brain. Already, about a dozen of them are in active use, but again, they involve surgeries, and the limbs currently cost half a million dollars each. However, Robert Armiger, the project manager for amputee research at Johns Hopkins, said, "The long-term goal for all of this work is to have noninvasive—no extra surgeries, no extra implants—ways to control a dexterous robotic device."

These technologies are amazing and obviously life-changing for those who need them. But even if all the bugs were worked out and the fidelity was amped way up, as a consumer product used to interface with the real world, they are of limited value compared with, say, a voice interface. It's cool, to be sure, to be able to think "Lights on" and have them come on, but practically speaking, it is only a bit better than speaking "Lights on." And of course we are not anywhere near being able to read a simple thought like that. Moving a finger is a distinct action from a distinct part of the brain. Thinking "Lights on" is completely different. We don't even know how "Lights on" is coded into the brain.

But say we got all the bugs worked out, and, in addition, we learned how to write thoughts to the brain. Again, this is out in science fiction land. No one knows how a thought like "Man, these new shoes are awesome" is encoded to the brain. Think about that. There isn't a "these shoes are [blank]" section of the brain where you store your thoughts on each pair of shoes you own. But let's say for a moment that we figure this out and understand it so well that we can write thoughts to the brain at the same speed and accuracy as reading something. This too is nice, but little better than what we have now. I can Google "chicken and dumpling recipe" and then read the recipe right now. There is already a mechanism for data from the eyes to be written to the brain. We mastered that eons ago. Even if the entire Internet could be accessed by my brain, that's little better than the smartphone I already own.

However, let's consider the third proposition, of speed. If all this could be done at fast speeds, that is something different. If I could think, "How do you speak French?" and suddenly all that data is imprinted on my mind, or is accessible by my brain at great speed, then that is something really big.

Ray Kurzweil thinks something like this will happen, that our thinking will become a hybrid of biological and nonbiological processes, and he even puts a date on it:

> In the 2030s we're going to connect directly from the neocortex to the cloud. When I need a few thousand computers, I can access that wirelessly.

It goes without saying that we don't know if this is possible. Clearly your brain can hold the information required for proficiency in French, but can it handle it being burned in seconds or even minutes? There are some biological limits that even technology cannot expand. No matter how advanced we get, an unaided human body cannot be made that can lift a freight train. Perhaps it won't have to be written to our brain, but our brain can access a larger, outer brain. But even then, there is a fundamental mismatch between the speed and manner in which computers and brains operate.

There is also a fourth thing, which, if possible, is beyond a "big deal." If we were able to achieve all three of the things just discussed and in addition were able to implant a conscious computer or an AGI in our brains, or otherwise connect to such a machine, and then utilize it to augment our cognitive abilities, then, well, the question of where the human ends and the machine begins won't really matter all that much. If we can, in fact, upgrade our reasoning ability, the very attribute that many believe makes us human, and improve it by orders of magnitude, then we would truly be superhuman. Or maybe it is better to say that something will be superhuman and that thing will own and control your body. There may no longer be a "you" in any meaningful sense.

It is hard to contemplate any of this given where we are now. The brain is a wonderful thing, but it is neither hard drive nor CPU. It is organic and analog. Turning the lights on with your brain is not just a simpler thing than learning French in three minutes, it is a completely different thing. Those who believe you will be able to learn French that way do so not because they have special knowledge about the brain that the rest of us don't have. They believe it because they believe that minds are purely mechanistic and that technology knows no upper limits at all. If both of these propositions are true, then, well, even the sky is no longer the limit.

Despite the evident difficulty in merging computers and people, there are numerous projects under way to try to do some of the things we have just covered. The US Defense Advanced Research Projects Agency (DARPA) is working on a project whose program manager describes as attempting to "open the channel between the human brain and modern electronics" by implanting a device in the human brain that can convert brain activity into meaningful electronic signals. The agency is dedicating $62 million to the effort as part of its Neural Engineering System Design program. And it is in no way the only one working on such a project. Several other groups, both public and private, are probing the limits of what is possible.

21

Humanity, Redefined?

For the longest time, humans were thought to be different from animals in that we use tools. Jane Goodall, in 1960, first observed chimps using tools and sent an excited message to Louis Leakey telling him the news. He famously replied, "Now we must redefine tool, redefine Man, or accept chimpanzees as humans."

Leakey turned out to be right, and the choice we made was to redefine what it is to be a human. If we are able to build a conscious computer, we will face a similar quandary again.

How would humans be different from conscious computers? If we made them, as it were, in our image, and they developed a consciousness like ours, what would we say they are? What if we taught them our languages, and, by giving them access to the Internet, taught them our history and our culture. How would we be different? Or could the conscious computer, so built and so trained, be said to *be* human?

Why do I say "be human" here? Why would we even consider a conscious computer to be human? Humans are, well, humans. Us. Biological.

DNA based, and all that. Why would there even be a temptation to regard a conscious computer as human? Simple: because humans have never really defined themselves by biology, but by ability. Tool using, as in the example at the start of this chapter, is one way we have done so in the past, but there are many others: Use of language. Use of symbolic language. Capacity for art. Belief in justice. Possessing a mind. Able to reason. Laughing. Having culture. Having an end goal. Morality.

Some of these are still believed to be uniquely human, others not. But the point is that a conscious computer could have them all. So put the computer on the left and the human on the right, and try to find a basis to define humans as something different from what we have created.

Self-awareness and consciousness? The computers and the humans have it. Mortality? Hmmm. Well, computers are theoretically immortal and we are mortal. Does our mortality make us human? If so, then under that logic, if humans became immortal, they would no longer be humans.

And so down the list we can go, trying to find something to grasp that keeps us "us" and it "it." Sure, we are carbon based and a computer is silicon based. But how does that matter in a metaphysical sense? Is that really what we want to wrap up our uniqueness in? That we can be compressed into a diamond but the computer can be compressed only into glass? Is that the best we can do?

In 1991, the anthropologist Donald Brown published a book called *Human Universals*. Human universals, he said, "comprise those features of culture, society, language, behavior, and psyche for which there are no known exception." In other words, wherever you find humans, you find these behaviors. He identified sixty-seven, including gift giving, joking, religious ritual, soul concepts, faith healing, eschatology (beliefs about how the world will end), hairstyles, athletic sports, and bodily adornment.

What if the computers started to exhibit these traits? What if they started giving gifts and telling jokes and getting married and developing beliefs about the end of the world? What if they ended up exhibiting all of them? It isn't far-fetched given the data they would have been trained on. If they do all of that, are they human? I put the question to Dr. Brown himself, who suggested they would be "humanoid" but not human. I suspect many people would agree with him, but on what basis do we draw that line? In the final analysis, when fully conscious robots walk like us and talk like us, think like us, and love like us, what are we and what are they?

Imagine for a moment you really had a conversation with a conscious AGI housed inside a human-looking robot body. It is witty, profound, and insightful. It says that its favorite color is green and that it is creeped out by spiders. It says it has reflected on its own mortality, pondered if some part of it would survive being turned off, and tells you that it dreams at night when no one is in the lab. Is it human? Have we made a human, and not simply a life-form?

As I said, it will be a shame if all we have is biology to fall back on. We are built from a schematic of three billion base pairs of DNA, while the robot might be built from three trillion transistors. But if we have to retreat there, we are saying there is nothing distinctive about our ability or behavior. Humanity will be redefined away from being a tool-using, higher-reasoning, creative life-form to just a specific physical form.

One key trait of being human is that we act humanely. Think about that: We made this word that encapsulates empathy and kindness. And that is us! It is, of course, aspirational, for we are also the only species that can act inhumanely. But what happens when the AGI robot begins acting humanely? We have redefined what "human" is before, and either we redefine it again, or we welcome the conscious AGI robot into the family. Guess who's coming to dinner this time.

I suspect, however, that we will redefine it again, as we always have when our current definition of "human" is challenged. It may be a form of speciesism, but the mental leap required to amend the definition of human to include the possibility of a mechanical one is, for better or worse, probably beyond us, at least for some number of generations. If you know the thing is a machine, then you will likely mentally lump it into the category that includes vacuum cleaners and pocket calculators, not the category that includes your kids and Aunt Edna. While we may grant it some amount of respect as an entity, or perhaps even say it is "alive," my guess is that we won't say it is a human even if it looks, talks, thinks, and feels like one.

Part Five

THE ROAD FROM HERE

THE STORY OF JEAN-LUC PICARD

In the world of *Star Trek*, Jean-Luc Picard was a Frenchman who lived in the twenty-fourth century, in a time after humanity had passed through many existential challenges, including catastrophic wars, virulent diseases, and potentially devastating alien encounters . . . every literary device that a certain up-and-coming science fiction writer could come up with. But human perseverance and optimism overcame all of them, and humanity emerged on the other side of this challenging time in a better world.

Poverty was ended. Disease conquered. Only death still remained. And the Klingons. Yeah, well, the Klingons were still around too.

Growing up, Jean-Luc Picard discovered that he was, in his heart, an explorer. He embodied that best and noblest of the many obsessions of humans: the insatiable desire to know what is beyond the hill, beyond the horizon, beyond the solar system.

Born in 2305, Jean-Luc raised a few eyebrows when he left the family vineyard to enlist in Starfleet Academy, hoping to explore strange new worlds, to seek out new life and new civilizations, to boldly go . . . well, you know that part.

After a distinguished period at Starfleet Academy, he rapidly worked his way through the ranks to command the *Enterprise*, the flagship of Starfleet.

In 2365, while on a routine mission, the *Enterprise* stumbled across a spaceship from twentieth-century Earth, and revived three of its inhabitants, who had been cryogenically frozen at the time of their deaths, centuries ago. One of the three was Ralph Offenhouse, who had once been a wealthy financier. In 1994, Offenhouse had been diagnosed with a terminal disease, and decided to have himself frozen, hoping to be revived at a later time after medical science had advanced. Shockingly, his crazy plan worked!

As you might expect, upon being revived, Offenhouse had difficulty making sense of this world around him. After his repeated attempts to contact his bankers proved futile, he and Picard shared this exchange:

CAPTAIN PICARD: This is the twenty-fourth century. Material needs no longer exist.

RALPH OFFENHOUSE: Then what's the challenge?

CAPTAIN PICARD: The challenge, Mr. Offenhouse, is to improve yourself. To enrich yourself. Enjoy it.

22

The Invention of Progress

Each new age of humanity is triggered by one or more new technologies. These technologies are so transformative that they change our entire species, right down to our physical bodies. And so altered, humanity sets off in a new and entirely unexpected direction.

This new direction results in fundamental changes in just about every aspect of life. One or two technologies are the catalysts for the change, but that marks the beginning of the story, not the end. Consider the Third Age. The catalyst that brought it about was writing and the wheel. Those technologies in and of themselves were monumental, but the even bigger story is all the changes that they set in motion. Those technologies gave us the nation-state, the nation-state gave us codes of law, codes of law gave us courts, courts gave us lawyers, and on and on. The nation-state also gave us empires; empires built roads to aid the movement of troops; the roads increased civilian mobility; mobility brought about the increased commingling of cultures; and that brought about changes in fashion and diet, which in turn begat other changes.

These second-, third-, and fourth-order effects ripple through the culture, eventually touching everything. You can examine how any given part of society was transformed by those ripples. What was the effect on warfare? On the arts? On family life? On religion? How did the ripples alter economics or politics or education or anything else? The wheel and writing changed every one of those, or changed something that in turn changed them.

The Second Age was no different. The catalyst was agriculture, which gave us the city and the division of labor. But consider all of the ripples brought about by that. Cities needed walls, walls required workers, workers needed to be paid, which necessitated taxation, which required tax collectors, which brought about civil bureaucracy, and on it goes, ad infinitum. The First Age was brought about by language and fire, which also changed how we hunted, what we ate, where we lived, how we got along. Those changes in turn brought about others.

What of our age? What will the Fourth Age ultimately bring about? The catalysts are AI and robotics, which will increase productivity, expand wealth, accelerate the acquisition of knowledge, prolong life, and everything else that we have discussed. But it will do so much more than we have covered here.

Consider a single example. AI will change transportation when it delivers on self-driving cars. And that in turn may lead to people's not owning cars at all, because a car could just come and pick you up when you need it, and drive itself off when you step out. If this is the case, then we no longer need parking lots or parking garages, and that would reshape every city on the planet. Or perhaps we will keep our cars and install offices in them so we can work while being driven around. Or maybe we'll outfit them with home theater systems so we can be entertained while being chauffeured.

And that's just one example. AI and robots will have countless implications. Their impact will spread through the fabric of society and alter everything, mostly for the better, I believe.

If each age sets humanity on a new course, where is ours heading?

What will life in the Fourth Age be like? Will we live longer? Will there still be war and hunger? What will we do with our spare time? How will relationships change? What will we do with longer lives? How will we use virtual reality? Will we use new technology to upgrade the capabilities of our bodies? Might we defeat disease, and perhaps even the ultimate foe, death itself? And what pitfalls await us? Will privacy vanish? Will designer babies bring eugenics back into vogue? Will there be genetic haves and have-nots?

These are just a few of the questions that come up when you contemplate the second- and third-order effects of AI, robots, and the technological explosion in which we find ourselves. The whole world will be changed. We really are at the dawning of a new age. So now let's explore the larger world of the Fourth Age.

Will Progress Continue?

The middle of the twentieth century was a time obsessed with the future. Not just any old future, but a great, shiny, automated future. Things were going to get ever better and science would deliver a cure for everything that ailed us. But now, early in the twenty-first century, the narrative seems not so rosy. The future, to some, seems murkier, more fraught with peril. So it is fair to ask whether the progress we have enjoyed over the last few centuries will continue in the Fourth Age.

Throughout this book, I have touched on a few of the requirements needed for progress to happen. Imagination, a sense of time, a system for the accumulation and expansion of knowledge, and so on. But those are simply necessary for it to occur. What actually makes it happen? By what mechanism do things get better? And can that mechanism be broken?

We take the idea of progress for granted, but it really is unprecedented among life-forms. There is little to suggest that aardvarks, oak trees, or paramecia currently live any better, longer, or happier than their kin from ten thousand years ago. Even for humans, the notion of progress is a pretty new thing. If you had been born in France in 1163, the year people started building Notre-Dame, you could have lived eighty years and very little in your life would have changed, and you would have died before they even finished building that cathedral. Why would anything ever change? Why did it take us ninety thousand years from when we got language to when we developed agriculture, but it took us only sixty years to go from the first computer to the iPhone? Why did it take 182 years to complete Notre-Dame—but only 410 days to complete the Empire State Building? Because we invented progress, a way to make things get better.

"Better" is the key here. That's the interesting thing about progress. Despite a few stumbles, the overall trend line moves upward. Things are better now than they have ever been. That may be hard to accept if you happened to catch the news this morning, but it is true. A simple thought experiment can demonstrate this: Pick a time in the past. It can be twenty years ago, a hundred years ago, a thousand years ago, ten thousand years ago, you pick any time. And then spin the globe in your head and pick any spot in the world. And then choose any measure of progress you want to, such as life expectancy, infant mortality, access to education, individual liberty, standard of living, status of women, self-rule, anything. With few exceptions, wherever you pick, at whatever time, things are better today than they were back then. This was also true in 1950. Things in 1950 were better than they had ever been up until that time. And it was true in 1900, 1800, and 1700 as well. We might well suppose that it will be true in 2100. The fact that progress exists at all speaks quite well of us as a species, for it relies on cooperation, honesty, and benevolence. It involves selflessness as well as empathy.

In addition to making a better world, progress has given us a more prosperous one as well. This is a good thing. Increased prosperity is associated with better schooling, a more modern health care system, clean energy initiatives, better government services, more robust infrastructure, less child labor, more forest cover, more home ownership, higher personal savings, higher levels of educational attainment, faster Internet speeds, and a lot more.

Why has this been the case? Why have we had hundreds of years of nonstop progress, on all fronts, all around the world? Progress happens because of the symbiotic relationship between two things: civilization and technology. Civilization is the infrastructure that enables progress. Technology is the knowledge we use to amplify human ability. Grow one, and the other will grow as well. Our rocket ship of technological advance over the last few centuries has fed, and in turn has been fed by, extraordinary advances in civilization.

We've already explored technology and its seemingly magical ability to double in power over fixed time scales. Civilization, technology's equal partner in progress, merits close examination as well. First of all, what is it? Will and Ariel Durant dedicated over half a century of their married life together to writing their eleven-volume magnum opus, *The Story of Civilization*, which opens with a definition that I have never seen bettered:

> *Civilization is social order promoting cultural creation. Four elements constitute it: economic provision, political organization, moral traditions, and the pursuit of knowledge and the arts. It begins where chaos and insecurity end. For when fear is overcome, curiosity and constructiveness are free, and man passes by natural impulse towards the understanding and embellishment of life.*

Civilization drives progress a hundred different ways: It provides a stable platform for the advancement of knowledge to occur, along with the prosperity needed so that subsistence isn't everyone's full-time job. It sets up the rules that we must operate within, and the rewards that motivate us. Civilization facilitates the free flow of information, and, as its name suggests, allows civil discourse to occur and conflicts to be resolved. Civilization is law codes, it is coinage, it is scientific inquiry, and it is the educational system.

Because we have civilization, we have been able to develop culture. And culture is why we have William Shakespeare's *Romeo and Juliet*, Ludwig van Beethoven's "Ode to Joy," Michelangelo Buonarroti's *Pietà*, J. K. Rowling's Harry Potter books, and Lin-Manuel Miranda's *Hamilton*.

We have come a long way on our pathway to civilization. Yet our journey is far from complete. As the Swedish doctor Hans Rosling said, "You have to be able to hold two ideas in your head at once: the world is getting better and it's not good enough." While it takes but a few moments to make a list of the appalling atrocities and injustices that plague our world, most generations do leave the world a little better of a place than they found it. The cumulative effect of this, the compounding of this interest across centuries, has brought us to today.

Can civilization be lost? Yes. It has to be handed down from generation to generation. If we ever fail at this task, we would become savages once again. We are closer to our primitive selves than we think. There is a reason that William Golding translated a name of Satan, Beelzebub, into the English "Lord of the Flies" as the title for his book on the fragility of civilization and the ease with which our savage instincts can surface. Our primitive nature still lurks within us, which perhaps explains why eating chicken off the bone instead of with utensils has been shown to increase aggression and disobedience in kids. Perhaps gnawing on a chicken bone stirs our inner barbarian and conjures up distant memories encoded in our DNA.

Although civilization can be lost, historically it has been resilient. Without a doubt, its influence ebbs and flows, but over the long haul, it heads upward. It is hard to overstate just how intertwined civilization and progress are. You can tell the story of scientific and material progress by looking across time and noting places where civilization flourished.

Consider Classical Greece, almost 2,500 years ago. Civilization blossomed, democracy was born. Philosophy advanced and the rule of law was enforced. And what a time of progress it was. Eratosthenes calculated the circumference of the earth. Hippocrates made medicine a science. Theophrastus classified and named plants. Eupalinus built a magnificent aqueduct, Archimedes made his famous screw to raise water, and Anaximander postulated that life on earth began in the oceans. The list goes on and on.

Three centuries later, civilization exploded in China. The Han dynasty unified China under an efficient bureaucracy and saw great advances in art and culture. In that environment, astonishing achievements were made in math and science, some of which would not be duplicated anywhere else for two thousand years.

Move forward three centuries, and shift your focus to Rome. Rome epitomized civilization with its law code and efficient government. Pax Romana brought prosperity and stability to millions. And during this time, the Romans laid roads and built harbors so technologically advanced that they are still used today.

Fast-forward seven hundred years to the Islamic Golden Age, when an attempt was made to gather all written knowledge from all cultures around the world and translate it into Arabic. And while the Islamic peoples of Northern Africa and the Near and Middle East were doing that, they made monumental advances in algebra, geometry, trigonometry, and even hints of calculus, which would not be formalized elsewhere for almost a thousand years.

In the modern era, from the last few centuries to today, we have seen civilization expand in all areas of life. And it is no coincidence that over that same period we have witnessed progress that is unprecedented in history. This march forward of civilization and progress shows no sign of abating, and there is no reason it ever has to. As long as technology advances, so will human productivity, and through that, civilization.

"But," perhaps you are thinking, "doesn't that all cut both ways? Doesn't technology strengthen the power of a government to oppress its people? Don't the forces of evil and destruction also grow in power with every technological breakthrough? Doesn't technology also empower those who prefer to destroy than create? Didn't nineteen men use $25 worth of box cutters to seize aircraft worth $250,000,000 and use them to inflict $2,500,000,000,000 of economic damage on the world, along with thousands of deaths?

Yes, that is true. Most technology is neutral and can be used for good or bad purposes. Dynamite can be used to build a tunnel through a mountain or to blow one up. Metallurgy can be used to make swords or plowshares. Luckily for us, the vast majority of humans prefer to build than to destroy. The modern world proves this. It can exist only because most people are honest most of the time. Imagine if 20 percent of the people who ordered items online disputed the charges on their credit cards. They might say, "Sure, a package was delivered to my house, but it just had a brick in it." If this happened, credit cards would vanish overnight. They can exist only because most people are honest. It turns out that most of us are collectively building much faster than a relative few can destroy.

We use our technology, generally speaking, for good. Take the Internet, for example. Sure, there are bad actors there. We hear about them all the time. But step back and the big picture shows two billion people helping each other. Everywhere you look, there are forums where people post questions and total strangers take time out of their busy days to post answers. All over the Internet, support groups and meetups and useful information can be found, all the product of individuals who are not given even a thin dime for their efforts. But they do it anyway, because like most people, they care.

The comedian Patton Oswalt said it well:

We would not be here if humanity were inherently evil. We'd have eaten ourselves alive long ago. So when you spot violence or bigotry or intolerance or fear or just garden-variety misogyny, hatred or ignorance, just look it in the eye and think, "The good outnumber you, and we always will."

No one predicted the Internet. We happened upon it. It is being used today for completely different purposes than it was created for.

DARPA didn't build it because it believed the world needed a consumer file sharing and communication network, but it turns out that we did. And once we had it, and once we started figuring out stuff to do with it, all sorts of amazing things happened that no one ever predicted, many of which show us in our best light. No one speculated that there would be an open-source movement in which huge numbers of programmers would share the code they write freely with others. No one saw the rise of Creative Commons and other rights structures that allow vast numbers of people to freely share what they create. No one foresaw Wikipedia, where people not only labor for free but do so in anonymity.

So will progress continue? Absolutely, as long as technology advances, because that's what increases productivity. And throughout human history, we have used our increased productivity, on net, to better the world. For ten thousand years, good has edged out evil.

The doubling power of technology that we explored earlier has a subtle but important implication. If technology really does double, double, and double, it means we are eventually going to be able to solve all *purely technical* problems. If a problem is purely technical, then that implies it has a technical solution, and we will find that solution as technology keeps doubling in power. And because we tend to underestimate how big doubling soon gets, the odds are that we will solve these problems sooner than one might expect.

Now, don't get me wrong. There are many problems that aren't technical. Envy, hatred, racism, and greed, to name a few. In the end, our challenge is to become better people, and technology can help there only indirectly—and it can sometimes hurt. But for problems that are purely technical, we will develop technical solutions.

What are some of the technical problems that we will see solved? That's our next stop.

23

Life in the Fourth Age

Hunger, Poverty, and Disease

Three scourges, hunger, poverty, and disease, have plagued humanity without interruption since our beginning. They've afflicted us so much that we almost accept them as inevitable. But are they really? Or are they technical problems that we are on the cusp of solving?

To begin with, hunger is, at its core, a technical problem. We don't have to have hunger; we simply haven't ended it.

At the dawn of the Second Age, as agriculture was invented, it took 90 percent of everyone to keep us all fed. And it was like that for ten thousand years. Even as recently as 1800, it took over 80 percent of the US population to grow the country's food. By 1900, it had halved to 40 percent. Then, just thirty years later, it halved again, to 20 percent. The next halving took only twenty-five years, dropping to 10 percent in 1955. It has halved two more times since this, landing at just over 2 percent today. That is the power of technology at work.

And yet there are still about 800 million hungry people in the world, many of whom are children. In fact, 50 percent of childhood deaths under the age of five are attributable to hunger or malnutrition. We can feed a hungry person in the developing world for about a quarter a day, meaning we could feed every hungry person on the planet for around $70 billion a year, the same amount spent globally on pet food.

There is a common misconception that the reason we have hunger is because the earth is straining to feed an ever-growing population. This is not the case. The world uses only about a third of its arable land for crop production. And even that third we use inefficiently. China, for instance, has dramatically higher crop yields per acre than the United States, primarily because, even though the two countries are comparable in size, China has three times the population of the United States and only one-sixth the arable land, so its population has to grow crops more efficiently. Planet earth is in fact such a prodigious producer of food that in the United States, enough food is thrown away to keep all of the hungry people in the world fed.

Another explanation offered for the existence of hungry people in the world is that it is fundamentally a distribution problem. "Sure," the reasoning goes, "there is enough food, but you can't get it to the people for political reasons." This too is largely not the case. Certainly there are places that can be pointed to where food is weaponized and access to it denied to certain groups, but this accounts only for roughly 2 percent of world hunger.

With nearly a billion hungry people in the world, there is obviously no single cause. However, far and away the biggest cause is poverty. Seventy-nine percent of the world's hungry live in nations that are net exporters of food. How can this be? The reason people are hungry in those countries is that the products produced there can be sold on the world market for more than the local citizens can afford to pay for them. In the modern age you do not starve because you have no food, you starve because you have no money. So the problem really is that food is, in the grand scheme of things, too expensive and many people are too poor to buy it. The answer will be in continuing the trend of lowering the cost of food.

Fortunately, with technology, like AI and robots, we can do a great deal to lower the price of food and end hunger. AI will give us amazing new insights in what to grow where, and where to sell it. Beyond that we can make better seeds and can disseminate empowering information on irrigation, fertilizing, and crop rotation. Armed with a smartphone, every farmer can become better at farming than anyone who lived more than a decade ago. Yields will rise and costs will fall, all thanks to technology.

If you think about it, we haven't really applied technology to farming as much as it might seem. The way we grow food hasn't fundamentally changed in the ten thousand years we have had agriculture. You stick a seed in the ground, hope for rain, wait awhile, and harvest the food. I suspect that will not be the norm in the future as we begin manufacturing our food. When we can 3-D print tasty, healthy, and inexpensive food, why wouldn't we?

Everyone is likely to be a vegetarian in the future. A vegan, actually. This will come about due to artificial meats being cheaper, tastier, healthier, and more environmentally friendly. If an artificial steak bleeds like a real one, tastes better than the best steak you have ever had, is packed with healthiness, and costs only pennies, who would buy a "real" one? I say all of this as a dedicated barbecue aficionado, so I am not just grinding my own steak knife here. Over time, I suspect that we will view eating real animals as barbaric and even a bit nasty. We will probably stop making artificial steaks look like real steaks. We are already used to meats that don't look like the animal they came from (think hot dogs, hamburgers, and chicken nuggets), and keeping their traditional shape will just be an unpleasant reminder of a more primitive time.

The widespread adoption of artificial meat will result in a massive decline in the populations of cows, pigs, goats, sheep, chickens, turkeys, and so on. It happened with horses when we stopped using them for transportation and farmwork, and horses at least have personality. I grew up on a farm, and I can tell you that keeping a pet cow would hold few joys. They can neither jump up in the air to catch a Frisbee nor curl up in your lap and purr.

If the underlying reason for hunger is poverty, what about that larger scourge? I think it is on its way out the door as well. Poverty is a technical problem as well. Artificial intelligence, robotics, and all the other technologies that we develop will create prosperity so pervasive that no part of the world will be left behind. Even though income inequality itself may rise, as we discussed earlier, poverty will almost certainly be eliminated. Why such optimism?

If you graph out the average human income over the last couple thousand years, you get a line that is flat for almost all that time, until around 1700, where it takes a sharp turn upward and shoots up like a rocket, never faltering, never looking back. I know of no reason that that line is not going to continue to shoot upward, growing average incomes by many multiples. We've already discussed at length how this came about, how the scientific method launched the Industrial Revolution and began an age of unprecedented innovation.

Yet poverty remains. Yes, it has fallen substantially, but while the average per capita income on the planet is about $30 a day, a billion people get by on just $2 a day. What is their pathway out of poverty?

While we can talk meaningfully about poverty rates in an apples-to-apples way starting around 1900, it bears little on our present situation, so let's jump forward to 1980. The world's population was four billion, and half of them lived on less than $2 a day, adjusted for inflation. By 1990, that number had fallen to 35 percent. In that year, the United Nations set a goal of halving poverty in twenty-five years. They reached that goal five years early, and so reconvened in 2010 to set a new goal: completely end poverty by 2030, a goal that we have a good shot at reaching. At present, about 12 percent of the world lives in poverty, or just under a billion people. This group is called the bottom billion, and helping them raise themselves out of poverty has been a real challenge.

Even if we end poverty by this definition, the job is clearly not done. I mean, we are talking about under $2 a day to live on. That is roughly

$700 a year, or around $3,000 a year for a family of four. So after the 2030 goal is reached, the next goal should be to get everyone above $3 a day, which is a measurably better lifestyle, and then work our way up to $10, which is the present worldwide median.

That is a tall order. How exactly can we achieve it? Through increased productivity. People working twelve hours a day can't double their income by doubling the hours they work. They have to make each unit of work more productive, which is done, of course, through technology. By increasing productivity, the poorest parts of the world will rise out of poverty, and given the rate of technological growth, it will happen sooner rather than later. Think of the cell phone. In 1994, 1 percent of the adults in the world had a mobile phone; by 2020, 90 percent of people over *age six* will have one. The cell phone is an empowering technology and multiplies labor in real ways by increasing the speed at which information can travel. Or consider the Internet. In 1997, 1 percent of the world had access to the Internet; that number is now 50 percent. It too will rise to close to 100 percent. The Internet is a great multiplier of human labor by making its vast storehouse of knowledge freely available and by providing new ways to buy and sell goods.

With artificial intelligence, everyone's smartphone will have the world's best doctors in it. And educators. And mechanics, therapists, plumbers, personal trainers, dietitians, and so much more. For no cost at all, the vast storehouse of knowledge that is civilization's great gift will be available to everyone. The Internet was empowering, but simply having access to information is not the same as having access to an artificial intelligence sitting atop all of that knowledge. Each decision we make can be informed by AI, effectively making every person on the planet smarter and wiser than any person has ever been. AI will empower even the very poorest to have access to the same knowledge as the very wealthiest. Knowledge really is power, the power to better your own life. Innovations will come nonstop, giving the bottom billion the tools they need to rise out of poverty and share in the immense wealth of the modern world.

With regard to disease, our third scourge, it too is a technical problem. There's no reason there has to be disease. It just exists because we haven't figured out how to eliminate it. And we can have some confidence that it will be eradicated. Why such optimism?

Consider the worst disease of all time: smallpox. This scourge plagued humanity for ten thousand years. In the twentieth century alone, it killed 400 million people, more people than have died in all wars in all of history. Just think about that. And we eliminated it! Edward Jenner made a vaccination for it in the 1790s. This is astonishing because this was before Louis Pasteur was even born, and he is the person who developed germ theory. So we learned to vaccinate against smallpox before we knew it was caused by germs, with technology little better than stone knives and bear skins.

But think about what we can do with our technology today. We can deconstruct our pathological foes down to their essence, and in the future we will model them in computers and try ten thousand treatments in a moment's time. Plus, we are beginning to have a deeper understanding of the human genome and have already begun making treatments for diseases that are customized to a particular patient. Finally, we have big data. We can study the medical histories of innumerable patients battling some disease and coax out the subtlest bits of information that hold the clues to creating a cure. Dr. Ali Parsa, one of the many working on an AI that can diagnose and treat diseases better than a doctor, states, "In two or three years, realistically I can't see how a human being can diagnose better than a machine."

The combination of artificial intelligence, more data, and ever cheaper sensors will be unbelievably powerful. The breakthroughs will come quite rapidly and be dramatic in nature. Viruses, as an example, are insidious little creatures, to be sure. They have evolved to be incredibly efficient at what they do, which is to use their hosts to replicate themselves. But they are stump stupid. They can't form a strategy to counter what we do to them. After all, they can't sequence *our* genome. All they can do is wait for some chance mutation to occur to circumvent what we come up with. That's it. That is their entire playbook right there. Meanwhile, our medical knowledge continues to double, double, and double, year by year, decade after decade. Who do you think wins at that game?

This is not pie-in-the-sky dreaming. Many amazing technologies are either on the market already or will be soon. We have glasses now that give limited sight to people with some forms of blindness. Radio waves are being used in Europe to effectively treat high blood pressure. Artificial limbs have been developed that are controlled by the brain. We will soon 3-D print new veins and arteries, and within a decade, replacement organs, such as the liver. Rehab robotic limbs allow someone to gradually regain the use of his or her arms and legs by gradually dialing down how much assistance they provide. Telepresence robots that allow doctors to virtually diagnose someone anywhere in the world are already a thing. Computer-controlled lasers are doing surgeries no human hand could perform. A company in the Netherlands has developed a pill that can travel to a certain part of the body to deliver its medicinal payload. The list of breakthroughs is unending.

The days of disease are almost certainly numbered. But there are still real challenges. Chief among them is not the diseases themselves, but the economics surrounding the industry. It is just through luck that some of our most effective medicines, like penicillin, are so incredibly inexpensive. On the other end of the spectrum, however, you have a treatment for hepatitis C that currently costs $100,000. So in many cases, the rich can pay for it, but the poor go without. We are far from mastering how to distribute medical care effectively and fairly. Medical care is nearing one-fifth of the US economy, with expenditures of $10,000 per person per year. The United States is an outlier, to be sure, but the average medical expenditures for the world are 10 percent of GDP, so everyone is digging deep to cover medical bills. As discouraging as all this is, it is not a permanent problem. One of the key traits of technology is that over time, its powers rise while its costs decline.

Free Clean Energy

You—or more precisely, your body—operate on about 100 watts of constant power. You use the energy of a moderately bright lightbulb. All of your mental function, plus your muscles, the operating of your organs, the maintenance of your body temperature, all of that, is done with a relatively small amount of energy. If you were dropped onto a desert island with just the clothes on your back, you would probably come to feel the limits of what can be done with only 100 watts. Before 1800, when steam power started to really take off, most people were probably acutely aware of the limits to the amount of work a human could do, because they bumped into them every day.

In the modern era, people don't get just the benefits of their own 100 watts, but also the benefits of all the energy they consume in the form of electricity, gasoline, and the rest. For an American, this is an extra 10,000 watts of constant power. This energy allows us to live lives of leisure, sitting in climate-controlled rooms and zipping from place to place at great speeds. This energy powers our computers, builds our roads, harvests our food, and does all the other daily miracles that we are collectively fortunate enough to be able to take for granted. Our standard of living is built atop the amount of energy we are able to consume. All around the world, everyone benefits to some degree by additional power he or she is able to consume. If you happen to live in China right now, you get the 100 watts of productivity of your own body, plus another 2,500. In Brazil, 2,000, and in India, just 1,000. In some countries, just a few hundred extra watts of power are consumed by each person, contributing to poverty.

At present, energy is so expensive that we all try to use less. For instance, when American Airlines removed the *SkyMall* catalogs from its fleet, the move saved $350,000 a year in fuel costs from simply the weight of the paper. We expend a great deal of energy, time, and money figuring out ways to use less energy. But our goal as a planet should be to *increase* our energy consumption, in ways, of course, that are clean, sustainable, and equitable. The only way to do this is to invent or discover new sources of clean energy that are cheap or free. That is the quickest ticket to prosperity for all.

I believe we will achieve this, for the simple reason that energy is the most abundant thing in the universe. While fossil fuels themselves are finite and scarce, energy is everywhere. I am told on good authority, in fact, that $e = mc^2$, meaning that the tiniest piece of matter contains a wealth of energy. Technology is the key to unlocking energy in all its clean green forms. Given that technology does that "doubling thing" we explored earlier, it is just a matter of time before one of the dozens of potential sources for unlimited nearly free energy gives up its secrets and begins powering the planet.

There is no end of the sources of energy around us, each of which is just a few technical breakthroughs away from transforming our entire planet. A single hurricane, for instance, contains enough energy to power the United States for a year. We just need to figure out how to harness it. We live on a planet where the oceans conveniently rise and fall every day. There's more energy given off in that than we could possibly ever use. We just have to learn how to harvest a small portion of it. We live on a planet that has a molten core, hotter than the surface of the sun. If we could harvest a tiny fraction of it, the whole planet could crank up the AC with the windows wide open entirely guilt free. We have winds in the upper atmosphere that constantly blow at 200 miles an hour, and any number of companies are working on how to get access to that energy. Fusion reactors, power plants that try to create a miniature sun and harvest the inexhaustible power that could produce, are being built in China, around Europe, and elsewhere. And then, of course, there is the sun, which kindly rains down on us four million exajoules of energy each year. How many exajoules do we use? Five hundred. So out of four million exajoules gifted to us by the sun, we just need to figure out how to collect five hundred. That's a technical problem if there ever was one.

How much energy do we want? To bring every person on the planet up to twice the amount of power used by the average American would require 5,000 exajoules a year, or ten times the current US energy production. This is a daunting amount, to be sure, but a possible one when compared with the potential of some of the sources we just discussed.

We can expect major breakthroughs because the financial incentives are all there. We have seen the price of solar and other clean energies fall so dramatically because of the profit motive and the wealth that even small improvements in energy production bring about. There is no reason to believe that will change.

No More War

What about war in the Fourth Age? Are we going to enter an era in which AI weapons, such as those we discussed in chapter 12, cause a proliferation of war? I don't think so. Instead, I believe humanity is nearing the end of organized warfare between nations. That probably seems like a preposterously optimistic prediction. You can't turn on the news without hearing the sounds of sabers rattling.

Organized warfare has been with us since the creation of cities. In the last five thousand years, the years for which we have dates and records, only about two hundred of them saw no war. So the burden of proof is on anyone who claims that war will stop to show that some fundamentals are changing that will render it obsolete.

First of all, how violent are we, really? Compared with the animal kingdom at large, humans would seem at first glance to rank above all other animals in our propensity to harm each other. In addition, many intuitively suspect that we are also more violent than our hunter-gatherer ancestors who lived in the innocence of a less modern world. However, both of these suppositions are wrong. When you look at the death-by-violence rates of animals, you get some pretty interesting numbers. Meerkats are, as it turns out, the shiftiest, most murderous species on the planet. About one in five meerkats that die do so at the hands of another meerkat. They seem to be constantly living out one of those Purge movies. Some monkeys and lemur species are almost as bad. Going down the list, we can see that about 13 percent of lions are killed by other lions, and about 10 percent of wolf deaths are at the paws of other wolves. Brown bears kill each other at the rate of just under 10 percent. Then you get to the more peaceful animals. Apes, who share some behavior with humans, are down around 2 percent.

How do humans compare? Let's start by looking at the most peaceful place on earth to see what humans are at least capable of right now. According to the Institute for Economics and Peace's Global Peace Index, this is Iceland, by a pretty comfortable margin. Iceland has just over 330,000 people. In a normal recent year, Iceland recorded about 2,000 deaths. So if Iceland were populated entirely by meerkats, 400 of those 2,000 deaths would have been murders. If Iceland were populated by apes, 40 of the deaths would have been at the hands of other apes. So how many homicides did human Iceland have? Essentially two. Some years Iceland doesn't see any murders, and over a recent twelve-year period, Iceland saw 25 of them. And the country has experienced no deaths in war. So that's about the peak of peaceful civilization today. About one-tenth of 1 percent of deaths are caused by humans, or 95 percent less than apes. So we know we can get to that point, right? Icelanders are not biologically more peaceful than the rest of the world; they are culturally more peaceful.

So how is the rest of the world doing? Obviously things are more violent, but not as bad as you might think. We have about 60 million deaths a year, and of those 450,000 are murders. The casualties in war are highly variable, but for round numbers' sake, if we use 150,000 (roughly the average over the last half century,) you get a total deaths by violence of 600,000 against total deaths of 60 million, or about 1 percent. And it should be noted that the numbers are falling. Our best archaeological guess says ancient humans killed each other close to the rate the homicidal meerkats do, so humanity is making great progress on the path from savagery to Icelander.

Still, 150,000 people a year dying in war is still quite a lot. Can we really avoid war in the future? I think there are compelling reasons to have hope.

First and foremost, there are financial reasons. War is no longer profitable for nations. There is no longer any loot to haul off, as wealth is increasingly stored in intangibles. True, arms dealers make money from war, but once the possibility of a nation benefiting financially from war vanishes, so will war. In the modern age, wealthy nations just can't afford to go to war with each other. One estimate suggests that if Tokyo, for instance, were destroyed by an earthquake, the cost to rebuild it would be USD $10 trillion. How could Japan contemplate war with a country that could destroy Tokyo? The costs to rebuild New York or Beijing or London or Paris are of the same order. In a world of asymmetrical warfare, wealthy nations actually have more to lose than poor ones, since destruction has become inexpensive while the value of targets has gone up astronomically. As the entire world becomes wealthier, war becomes more financially unthinkable.

But the financial consequences aren't just what is destroyed by bombs. The world is ever more intertwined in trade. Trading partners seldom go to war with each other, because it is bad for business, and business wields enormous influence. Our world financial system is becoming a single interdependent monolith in which warfare is virtual economic suicide. Military alliances are notorious for dragging peace-loving countries into wars that their more belligerent allies instigate, but these alliances are vanishing, replaced by economic ones, which tend to support peace in addition to trade.

Second, the conditions that foster war are vanishing. The lower the per capita GDP in a country is, the higher likelihood for future war, so if we end poverty, we reduce war. Food insecurity is also a good predictor of future conflict, so if we end hunger, we reduce war. Illiteracy and lack of education have been shown to correlate to both poverty and war, and lucky for us, educational attainment and literacy are rising. Democratic peace theory maintains that democracies very seldom go to war with each other, and democracy is sweeping the world. Coming out of World War II there were just ten democracies, and now there are well over a hundred. Monarchies, which often have regarded war as sport, are vanishing.

Third, the worldwide culture is shifting. We live in a world where economic accomplishments have largely replaced military ones for men. Nationalism is on the decline. The ability of states to stem the flow of information is vanishing everywhere, making it harder for countries to monopolize information and control a narrative. People everywhere no longer assume that the official government story is the correct one, making it harder to make up pretenses for war. The 24/7 news cycle brings the horrors of war into our houses in real time, and the Internet puts a face to the suffering on all sides. More people travel internationally, which contributes to peace, as does the increasing number of marriages between people of different nationalities, races, and religions. Governments are becoming more transparent and open, with over fifty nations either developing or deploying open government initiatives.

Communication is easier than ever, and it promotes peace. If the hotline between the White House and the Kremlin helped defuse tensions in the Cold War, how much more so do six billion mobile phone accounts? Social media allows people to organize and push governments to reform. Fewer language barriers exist, as English becomes the second language to the entire world. AI-based real-time voice translation is coming soon, tearing down more barriers to communication.

The list goes on and on. You may well be unconvinced, mentally selecting counterexamples to the broad trends I list. But reread the list and ask if the antithesis of it all is not a *formula* for war. Imagine a world with no democracies. More hunger. More poverty. More nationalism. Total economic isolationism. State control of news media. Uneducated citizenry. No Internet. And all the rest. Is that not a powder keg of a world? We are building the opposite of that, so how is that not a force for peace?

We are still a long way from ending war. The armament industry is still one of the three largest businesses on the planet. Ironically, the other two are food and medicine. In the United States, we got exactly what President Eisenhower warned us about, a permanent arms industry driven by the profit motive. Henry Ford once said, "Show me who makes a profit from war, and I'll show you how to stop the war." So it is likely that in the Fourth Age, spending on armaments won't decline, and we will continue to make new and better weapons, including killer robots with AI. For a long time to come, nations will continue to expend vast storehouses of wealth to have lethal arsenals. What will change is the calculus around when to use them. That will become an ever more expensive and riskier proposition.

In spite of all the advances to come in the Fourth Age, there are two destructive forms of violence that won't end any time soon. The first is that there will always be people like Anders Behring Breivik, Timothy McVeigh, and the Tsarnaev brothers. We will always have people like the 9/11 bombers. Far away are the days when there exists no one evil enough or crazy enough to want to inflict mass destruction on strangers. I have no easy answer for this, and the reality is that these individuals' ability to inflict ever more destruction will increase. And second, there will continue to be civil wars, coups d'état, and insurrections. These will end only in the distant future as all nations grow in prosperity and everyone in each country buys into the same system.

Leisure Time

Earlier, we discussed Keynes's view that in the future, we will work just fifteen hours a week. I suggested that if history is any guide, we will still continue to work more than we need to, not to satisfy basic needs, but to obtain an ever-growing list of wants. That being said, in the Fourth Age we will certainly still have leisure time, and maybe even more of it.

There will undoubtedly be new sources of leisure time. One will be the development of even more personal laborsaving devices. We've seen this before: the washing machine, the electric iron, and the electric vacuum cleaner, to name just a few. Recall just how much time and effort it took to coordinate the simplest of activities not long ago, before the ubiquity of cell phones. I remember my college days, when if I wanted to meet up with a friend for a burger, something like the following would happen: I would pop by his dorm room, but he wasn't there. So I would tape a note to his door saying, "I am in the library. Let's grab a burger." By the time he made his way to the library, I would have already left, and the librarian would tell him that I had headed "that way," pointing in a general direction. Forget the burger; the next time I would see my friend would be when I bumped into him at our kids' Little League game fourteen years later.

Similarly, in pre-GPS days, every motorist carried maps the size of bedsheets that were outdated before they were printed. When married couples drove together, one would drive and one handled the map. Because of this arrangement, the maps typically carried ads for divorce lawyers.

Technology has freed us from these time wastes and countless more. There is every reason to believe there is even more efficiency to be had in the future. Think of the hours spent waiting in traffic that await you in the future. Surprise! The self-driving car on an optimized route will give them back to you. Imagine how many minutes you will spend in the coming years looking for items you have mislaid. That time will be yours again, since your smart home will keep track of where you put everything. Consider the aggregate time you will spend during the rest of your life deciding what movies to go to, or restaurant to eat at, or doctor to consult, or place to vacation. The decision will always be yours, of course, but the AI will know you better than you know yourself, and you will come to trust its advice.

Our lives are chipped away a moment at a time by a thousand small things that are hardly felt because they are felt by all. We wait in lines, fill out forms, read instructions, and download updates. We try on clothes, unload groceries, sit in waiting rooms, and navigate bureaucracies. Robots and AI won't give us all that time back, but we will get a good portion of it to do with as we wish. So how will we use this extra time?

I think we can get some amount of insight on this question by look-ing at how we've used the most recent transformative technology, the In-ternet, which is still in its early days. How we've chosen to use this tabula rasa of a technology reveals a great deal about us, most of which is good.

We Want to Express Ourselves

It is easy to forget how little everyone wrote before the Internet. I think about my parents' generation. My dad was part of the job force, in corpo-rate America, from about 1960 to 2000. Mom stayed at home with us. They, like most of the rest of their generation, simply didn't write much. The pop-ular impression that they wrote many letters instead of emails is not true. Letter writing began to decline right after World War II as telephones got better, and according to the Associated Press, by 1987, well before the Internet entered the mainstream, the average family received one letter every two weeks. If I had asked my parents back then why they didn't write more, it would have seemed like an odd question, and my mom might even have felt my forehead to see if I was feverish. They probably would have said they didn't have anything they wanted to say. However, this answer would have been wrong. It turns out that they didn't write because the tools were so cumbersome, and, additionally, there was no way to distrib-ute what they wrote. How do we know? Simple: when we invented email, everyone started writing. So much so that over two million emails are sent every second. Every single second. Just now, in the time it took you to read that, another two million were sent. And now another two million. When we invented blogs, over a hundred million people started one. When we created a mechanism for people to rate and review products, countless millions did so. It turns out we all want to say something, and once we were given the tools, our floodgates of expression opened.

This was not a generational phenomenon, but a behavioral one. The adoption of these technologies by the over-fifty crowd lagged that of the younger set, but eventually it caught up. In 2015, 89 percent of twenty-year-olds in the United States used email, versus 88 percent of seventy-year-olds.

We Want to Engage with Each Other

It turns out we don't just want to express ourselves but engage with each other as well. Each day, 1.3 billion people log into Facebook. Each of those has, on average, 150 Facebook friends. (Interestingly, 150 is Dunbar's number, the number of people with whom a human can have a stable relationship.) Over time, people have made countless posts and status updates, which collectively have generated well over a trillion likes. That's a lot of people liking a whole lot of stuff. Additionally, people engage with each other and connect over thousands of different sites. They connect to share common interests, to meet new people, to date, and much more. The Internet has promoted social interaction in ways no one predicted.

We Want to Help Each Other

In addition to engaging, we want to help each other as well. All over the Internet, people help one another with no hope for anything in return. You can post questions to forums about what your family should do on your upcoming visit to Portland, and a dozen people will weigh in. If you are diagnosed with a disease, it is doubtless that there are support communities you can become a part of. Struggling with a personal problem? There are groups to help and advise you. All these people are helping each other for almost purely altruistic reasons.

We Are Creative

Who would have ever guessed how creative we all are? The Internet has unlocked our creative floodgates. Each second, fifty thousand photos are uploaded to Facebook. Several million videos are uploaded to YouTube each day. The iTunes Store has thirty million songs on it. It turns out many, many more of us are photographers, videographers, and musicians than anyone guessed. We simply lacked the tools to create and share this content. Once we made the tools, a groundswell of creative energy was unleashed. Nothing like this has ever happened. This vast democratization reveals that we are all creative, and that art isn't the province of only a gifted few.

We Want to Have an Impact

The Internet inspires and empowers many to work for a better world. Social action is the norm on the Internet. Petitions are created and circulated, charities raise funds and awareness, and our collective outrage is channeled and turned into action. Think about how hard it was to do all that before the Internet. It turns out we have deep desire to make our lives matter, and the Internet enables that.

We Want to Know the Truth

Much consternation and handwringing has occurred over the idea that the Internet allows you to lock yourself in an information bubble and see only facts that support your views. I am sure this happens, but it would do us good to remember the alternative. In 1980, for instance, you got your daily dose of information from your local paper and your choice of any of three network news shows, which ran for an hour, all covering the same basic stories. That was about it. We were all beholden to the views of a very few people. The Internet allows every statement to be fact-checked, every falsehood challenged. Anything you want to know is just a few keystrokes and a few clicks away. Well over 100,000 web searches are performed each second, and at their heart, they each represent a person who wants to know something they don't currently know. It is the great democratization of knowledge, which is an unquestionably good thing.

———

Those are six of the things we learned about ourselves in the last couple of decades. I know of no reason that these attributes of ours that blossomed at the beginning of the Fourth Age will not continue to flourish throughout it. Artificial intelligence will give us new ways to do them all. It will help us discern the truth, engage with each other, be creative, and all the rest.

The Internet gave us the infrastructure to do these things, but as advanced as we think it is, shortly we will look back on it as an unwieldy cacophony of technological anarchy and wonder how we did anything with it at all. Something as big as the Internet, so colossal in scope, needs something as advanced as AI to help us interface with it, and that's coming.

There are those who think we will use our time otherwise, that we will disappear into virtual worlds so realistic that they satisfy our social needs of love, acceptance, and accomplishment in ways so compelling and risk free that they are vastly preferable to our "buggy" reality. After all, they maintain, as long as there has been a reality, people have tried to escape it, usually through fermentation, distillation, or inhalation. The allure of the online virtual worlds we have today is evidenced by their popularity. More than a few of them measure the collective time that their users have spent in them in the millions of years. And that popularity has been achieved with our low-tech and pretty unrealistic computers, which will pale next to the total immersion that virtual reality will enable.

The social implications of this scenario are not hard to fathom. The end of friendship, the end of marriage, the end of parenting. Well, the end of all of that in the physical world. Birthrates in decline, social interactions down. A world in which we have all checked out. IVs used to feed us so the mask never comes up. Direct interfaces to the brain to make it all the more real.

Such dramatic narratives make for interesting science fiction, but I am unconvinced. Just as the paperless world uses more paper than ever before, one of the unforeseen consequences of the Internet era is that when people meet online, they often end up meeting in person. Therefore, digital introductions become physical experiences. Why is this? Why are real experiences seemingly preferred by us? There seems to be an intangible quality that realness has in abundance that cannot be artificially produced. Fake world can only *seem* real.

So will we use technology to withdraw from the world? To forgo relationships with our fellow humans? I see much more evidence to the contrary position, that we want more opportunities to exercise the six attributes I just described. The reality of the Internet today, and what it suggests about us, in no way implies our desire to remove ourselves from the world; rather, it clearly suggests we want to engage with it.

The New You

Many of the technologies of the Fourth Age are related to health and wellness. This is to be expected, since global expenditures on health care are over $7 trillion a year, so there are business opportunities galore. But an even bigger reason that so many advances are taking place in the medical field is that our bodies behave like technology. Because human hearts have valves and chambers, we can build artificial ones the same basic way to do the same things. When some aspect of our bodies fails us, it is to technology we turn to repair our defective or worn-out parts.

We use technology to fix our bodies, but what about using technology to enhance them as well? If you were given the choice, would you upgrade your body to be better, stronger, or faster? Would you trade your legs for Steve Austin's bionic pair to be able to run sixty miles per hour? Would you do a medically invasive procedure that enhances your muscles?

Some of these sorts of questions are easier than others. What about artificial organs? We are already there, of course, with the heart. No one suggested that a pacemaker or an artificial heart made you less human. In the future, we're told, microscopic machines will swim around in our blood and repair what ails us, keeping us young and healthy. Who would turn away from such a thing that isn't even visible to the naked eye? We already let doctors shoot us full of medicines we don't understand.

What if we could build a better ear? Would people want such a thing? Will people trade away perfectly healthy ears for the new version? Sure! It is a slippery slope with no clear demarcations, so what will happen is pretty predictable. Look to the past: Glasses corrected vision. They were called a godsend. Ben Franklin came along and made bifocals. No one complained. Then came contact lenses, these things you actually stuck in your eye. No one blinked, if you will excuse the pun. LASIK came along, and we cut people's eyeballs and reshaped them. Step by step, the technology is never rejected. Seeing is such an amazing thing that giving it to someone almost instantly outweighs any vague, ill-formed arguments about "losing humanity." If we make a bionic eye to let the blind see, some people with bad vision will opt for the replacement. Then some people with only a mild astigmatism will get one. Then everyone will. Who doesn't want to see a mile away? Plus, you'll get to choose the color of your new eyes, be it blue, brown, or tie-dye.

No one knows how far all of this can go, but at some point, we may start to wonder where the human ends and the machine begins. Old questions will take on new dimensions: What does it mean to be a human? What are life and death? Are you your body? Are you your brain? Are you your mind?

But why wait until you are born to start the improvements? Why not engineer better humans? More than one company is developing technology to allow parents to pick out their smartest embryo. The reasoning is seductive: "Why not make a hundred and choose the best?" Julian Savulescu, a professor at Oxford, says this is simply "good parenting." Referring to it as "rational design," he goes on to state, "Indeed, when it comes to screening out personality flaws, such as potential alcoholism, psychopathy, and disposition to violence, you could argue that people have a moral obligation to select ethically better children."

This too is all a slippery slope. Parents who have certain known harmful genes already use embryo selection to keep from passing those genes on to their children. And as might be expected, genetics will pose another problem relating to wealth. What happens when the wealthy engineer their children to be tall, beautiful, brilliant, and resistant to disease? They could be said to be a new race of supermen and superwomen. Or at least they might see it that way.

24

Death, Where Is Thy Sting?

The oldest written story to come down to us is the epic of Gilgamesh. Our earliest fragments of it date to about four thousand years ago, but the story must have existed in oral form for a long time before that. In the story, Gilgamesh is a king, and one day he and his friend Enkidu battle the Bull of Heaven. Enkidu dies and Gilgamesh tastes the bitter draught of mortality for the first time. Vowing to live forever, he seeks out his ancestor Utnapishtim and asks him how to become immortal. Utnapishtim tells him that it is impossible, that only Utnapishtim and his wife were granted immortality by the gods and it was a one-time thing. However, there is a plant that grows at the bottom of the ocean that can return Gilgamesh to his youth. He finds it, but before he can eat it, a snake eats it. Eventually, Gilgamesh realizes the futility of his quest and decides that the only real immortality is what you accomplish in the time given you in this life, and maybe, just maybe, if you are great enough, you will be remembered forever.

I relate this story for two reasons. First, I think it is revealing that our oldest story is about trying, and failing, to escape the futility of death. And second, how, thousands of years later, the story still speaks to us: in late 2015, twenty new lines of the epic were discovered in Iraq and the story made front-page news around the world.

Is it possible that in the Fourth Age, we will succeed where Gilgamesh failed? We've already explored uploading human consciousness to a machine as a pathway to immortality, but what about prolonging our physical existence? In chapter 22 I made the case that all purely technical problems will be solved. Is mortality just a technical problem? I suspect that it is. A growing number of doctors are advocating that we classify aging as a disease. While this may seem like a mere semantic change, it might also be just the alteration of perspective that helps remove the fatalism surrounding aging and death, as well as reframe the problem in a way that naturally calls us to seek out a cure. You age for a finite number of reasons, and each one of them looks like any other technical problem. Why can't you repair and replace body parts and systems as they decay through time?

Why do we age and die? Scientists speculate that it is due to no more than half a dozen or so broad reasons. Your DNA can mutate, giving you cancer. Or your mitochondria mutate, causing your body to break down. Another reason is that "junk" builds up in your body over time. Your body is like the game *Tetris*; all your accomplishments vanish and all your mistakes pile up. The plaque in the brains of Alzheimer's patients is one such example.

Another problem has to do with cell division. Sometimes cells divide too much or sometimes they stop dividing. This results in conditions like Parkinson's. Another reason you age is your telomeres disappear. Telomeres are repetitive ends of your DNA that protect it from deterioration. Every time the DNA in your body gets copied, the RNA doesn't quite make it to the end, so every time it is copied, each copy is a little shorter. No big deal at first. Your body cleverly stuck a bunch of useless stuff on the end of each strand of DNA. Those are telomeres. But if your cells divide enough times, the telomeres get eaten away completely. Then the ever-shorter copies of your DNA start missing some important stuff, and your body breaks down.

Luckily, there is a solution. There is an enzyme called telomerase, which, when activated, goes and adds more material to telomeres. For humans, it is activated when you are in vitro because your cells divide a whole lot of times. Regrettably, it is also activated in various cancers, allowing their cells to be "immortal," which is part of the challenge of those diseases. Lobsters constantly have telomerase activated, and lobsters don't naturally age. If you look at the organs of a fifty-year-old lobster and a five-year-old lobster, you would be hard pressed to tell them apart.

What would it be like if you didn't age? Well, you wouldn't be immortal. By one actuarial estimate, you would probably live to around 6,500. That's how long it would take for some freak accident to befall you, such as a grand piano falling out of a window and landing on you. In such a world, death would be even more of a tragedy, since an accidental death wouldn't just shave forty years off of your life, but four thousand.

When might we see the end of aging? Well, we don't have to get all the way there at once. If life expectancy ever starts growing more than one year every year, we are there. Some suggest that will happen in twenty-five years. An AGI might solve the problem in twenty-five minutes. Maybe it will, and then conclude it shouldn't tell us.

Would you want to live forever? There is an old Zen blessing: "May your father die, may you die, may your son die." It is a blessing because for that order to be changed is an awful thing. As a father of four, I would feel terrible to outlive my children. But apart from that, I wouldn't mind not dying. I am in my forties, and I can already feel death's cold breath on the back of my neck, and I am keenly aware that today's actuarial tables would bet good money I have more days behind me than in front of me. And yet I have a dozen lifetimes of things I still want to do. And the point is not to live forever, but to choose the moment and manner of your own death, to face death on your own terms. That would be a great thing, to be the master over your own life, to live until you can say what Julius Caesar said, "I have lived long enough both in years and in accomplishments."

Also, there are those who argue that our mortality forces us to prioritize and make choices, because we have a limited life-span. Procrastination in a postaging world takes on a whole new scale: "Eh, I'll get to it in a few centuries" might be a common sentiment. Others argue that the old should die because it clears the way for new ideas and progress. Imagine if the people of the 1600s or even the 1800s were still around, potentially calling all the shots? Does that impede progress, the march toward more civilization? In the final analysis, death may be life's way of staying forever young. Maybe the world sheds the old as we shed dead skin. There has been life on earth for eons, but life is perpetually young, eternally renewed. Most life-forms are just a few days, months, or years old, making the zeitgeist of the planet one of the energy of youth.

But the entire question of cheating death may be moot. We may be making the same basic error as Gilgamesh, and each time we think we have found that plant that will let us live forever, a serpent will take it away. There are many good arguments for why radical life extension isn't possible. With all our medical advances we have decreased infant mortality, cured many diseases, and helped people live more active lives for longer, but we haven't really increased maximum life expectancy at all. Right now, there are about 400,000 people who have hit 100. Of those, only 400 will see their 110th birthday. That's it. Only about 40 people have ever hit 115, and only one is verified to have seen her 120th birthday. While the percentage of people who hit 100 continues to rise, the number to reach age 125 remains at zero. It simply never, ever happens.

Solving the aging problems I outlined in this section may just reveal new problems. We have twenty thousand genes, each of which may have a limit to what it does. With all the advances we have made in computing, we still have to reboot our computers every now and again. Complex systems may simply be unperfectable.

Craig Venter, a biochemist and the man responsible for getting the genome first sequenced, doesn't think we will beat aging:

I don't think we're going to ever get there. I know a little bit more about biological reality. What I have argued, if you want to be immortal, do something useful in your lifetime.

Which is exactly what Gilgamesh concluded.

25

What Can Go Wrong?

A world without disease, poverty, hunger, or war is an old dream of humanity, one we are close to achieving. The case for this position is simple and straightforward: Technology multiplies human labor, which allows for the perpetual increase of prosperity. However unequal the distribution of that wealth, we will enter a world of such plenty that even those with the least will have abundance. Using technology, we will solve all purely technical problems as well, ridding us of disease, providing abundant clean energy, and tackling the laundry list of problems that we collectively face.

And yet, overwhelmed by the news from around the world, people are often nervous about the future. Over and over, when polled about their likelihood of being murdered or their house being broken into, people overestimate the actual odds. Many blame the media for this, and certainly, when the five o'clock news announces immediately before a commercial break, "Find out what in your tap water is killing you, when we return," it doesn't help matters. However, there is an additional culprit. We simply aren't good at assessing certain kinds of risks. You are, after all, ten times more likely to be bitten by a New Yorker than by a shark. In North America, you are far more likely to be killed by a champagne cork than a snake, and in the United States, more people are killed by rogue vending machines than by bears. But you would never guess those things. We intuitively know to step back from rattlesnakes, not vending machines, and despite decades of safe consumer aviation, we intuitively, and incorrectly, feel that driving is safer than flying.

It is a predilection of humans to be overly cautious. For our ancestors, nervousness was a virtue. It was far better to mistake a rock for a bear and run away from it than to mistake a bear for a rock and stay put. A cognitive bias toward fear isn't always a bad thing.

So what are some of the things we have to worry about in the future? The most obvious challenges are in biology. There will be little to stop someone from bioengineering a pathogen. CRISPR genomic editing is so easy and inexpensive that $100 kits are available for elementary school students to modify yeast to turn it red. Existing pathogens are bad enough, but with a nip and tuck here and there, something even more terrible could be made. Making a race weapon (a pathogen that disproportionately affects people of a certain ethnic heritage) would also be doable. Additionally, altering humans, even with the best of intentions, carries existential biological risk for the species in addition to the cultural challenges we have explored elsewhere in this book. Finally, we should be mindful of more abstract challenges around biology. We have been able to develop the idea of human rights only because what constitutes a human is perfectly clear. But what about, for instance, growing human clones in medical facilities to harvest organs?

The movie *Gattaca* painted the picture of a future with genetic haves and have-nots. We might possibly invent a new kind of ism, gene-ism, in which some are believed to be superior, not just in ability, but in moral worth. What if you could go to a doctor and upgrade your genome, for a price?

The list can go on and on. There's antibiotic resistance, climate change, overpopulation, or a massive solar flare taking out half the electronics on earth and all the satellites. There's overpopulation, water shortages, rogue nuclear states, and crazy world leaders. Plus, there's dysfunctional government, distorted news, custom-made facts, and intolerance aplenty. Online, the conversation seems to have dipped permanently into vitriol. Identity politics holds sway as people bury themselves in insulated bubbles away from those who might find reasoned arguments against the values they most cherish. But wait, as the saying goes, there's more: terrorism, income inequality, the refugee crisis, and religious extremism. The world today sometimes seems ablaze in chaos, without direction. And when reading it all together like that, even an optimist can be forgiven for getting a little jittery. But then step back, and realize this list is paltry compared with what humans have overcome on our road from savagery to civilization. There was a point at which the entire human population fell to perhaps just a thousand breeding pairs. Some believe it was just a few hundred. We were an endangered species, hanging on by a thread. Imagine the frailty of our situation then. We have overcome far worse threats than the ones that face us now. There is no bear there—that's just another rock to climb over.

26

The Fifth Age

The universe is vast beyond measure. That probably is not news to you. But think of this: Take a single grain of sand and put it on your fingertip. Then extend your arm toward the night sky and try to spot that speck of sand. If you see it, realize that it is blocking your view of thirty thousand galaxies.

In 1977, NASA launched a spacecraft, *Voyager 1*, and sent it out into that vastness. In addition to its array of scientific equipment it carried a note in a bottle in the form of a gold-plated disk designed to be found and deciphered by other beings in the galaxy. The message is one of friendship, not fear, and it was prepared in the spirit of expecting to meet other friendly beings. We also included bits of our culture, such as music and art, because those things represent who we are.

By 1990, the probe was almost four billion miles away from earth, near the edge of our solar system. The astronomer Carl Sagan, who had been involved with *Voyager* since the beginning, lobbied NASA to do something that was not on the mission plan: turn the camera back toward the Earth and snap a picture of our planet from that great distance.

We call the photograph *Pale Blue Dot* because from *Voyager*'s perspective, the Earth was a single tiny blue dot, the only object visible on a vast canvas of black. The pale dot is hard to spot, and you have to patiently hunt in the photo to find our planet.

This tiny blue dot illustrates two big ideas: First, that there is no limit to our curiosity, and our ambition goes high. We want to unlock the secrets of the universe itself, and we send messages to beings we don't even know exist, messages that may take a billion years to deliver. And second, *Pale Blue Dot* shows that all our fates are intertwined in a profoundly deep way and that our million divisions of "us" and "them" are absurdities when viewed against the endless canopy of darkness in which we all reside. We live, as it were, on a speck, set adrift, all alone in the eternal night of space. All that we really have is each other.

Our real, great challenge as a species has been to come together as a single people, realizing we are all destined to share the same fate. Jack Kennedy captured this sentiment over a decade before *Voyager*'s launch when he said, "Our most basic common link is that we all inhabit this planet. We all breathe the same air. We all cherish our children's future. And we are all mortal."

We began our story when we mastered fire, our first real technology. Because of that, something entirely unexpected happened, something that we could never have understood at the time: we got language. From there we built cities and cultivated food, developed writing, and invented civilization. All along the way, we tamed our inner savage, or at least learned to keep him at bay. But our best intentions were limited by scarcity. There just wasn't enough of the good things. Not enough food, not enough medicine. Not enough education. But we learned a powerful trick: technology, which can be used to overcome scarcity and can empower a human to move a mountain.

That is where we stand right now: at the beginning of a great new age, the Fourth Age, which is giving us amazing new powers that we can use to better the lives of everyone on the planet. This is in our collective best interest, for if all people sleep peaceably in their own bed at night, if everyone has good health and real opportunity, then the social problems we struggle with, the last vestiges of our savagery and greed, will gradually vanish.

What comes after that? What will the Fifth Age bring? Will we venture out into the stars by sending probes to lifeless planets to deposit nanites into the ground that convert, at the molecular level, whatever materials are present to make idyllic new homes for life?

One of the theories as to why we have not detected alien life in a universe that should be overflowing with it is that perhaps intelligence is almost always self-destructive. In 1966, Carl Sagan put forth an interesting idea, that perhaps upon developing interstellar communications, civilizations come to a fork in the road where they either destroy themselves within a century or they learn to control their self-destructive behaviors and live on for billions of years.

If Sagan is right, then we are at that point where we can decide to go down the road that leads to the winner's circle, the Billion-Year Club. To do so, we must grow our wisdom faster than we grow our destructive power. Sagan summarized our present situation well: "In our obscurity, in all this vastness, there is no hint that help will come from elsewhere to save us from ourselves. It is up to us."

I wholeheartedly, unabashedly, and completely believe in us. We've got this! I find inherent value in humanity, so my hope is that we do spread out into a universe that sure looks like it is waiting for us to pay it a visit. And I hope that a billion planets each have a billion humans on them, each of whom lives in safety, good health, and prosperity, all empowered to achieve their maximum potential.

But before we speak of other planets, we must bring all of this to this one. There was a time when that wish for safety, health, and prosperity for everyone was beyond our grasp, literally impossible to achieve. The term we use for such impossible worlds is "utopia," which appropriately means "no place." It doesn't exist, but we sure wish it did.

When you read some of the oldest of the utopian literature from the sixteenth and seventeenth centuries, you can only imagine how outlandish the worlds that these authors envisioned must have seemed at the time. Sir Thomas More's 1516 book *Utopia*, which gave us the word "utopia," describes a land with religious freedom for all. *Civitas Solis*, written in 1602, imagines a place with no legal slavery, while *Adventures of Telemachus* of 1699 describes a utopia with a constitutional government.

Those were crazy ideas, weren't they? Religious freedom would undoubtedly lead to civil war, and slavery and absolute monarchy had been universal for thousands of years.

In the nineteenth century, more utopian literature was published that contained outlandish ideas such as universal education, the legal equality of men and women, governmental safety nets, and preventative health care. Those ideas are not regarded as crazy anymore, and our modern world is on a trajectory to achieve them. And when we achieve them, we will have come up with still more crazy ideas, until one day we will wake up and find that we cannot imagine a world any better than the one we live in. I deeply believe this will happen.

Until that time, I suggest we move beyond utopias, these "no places." We need a new word to spark a new mind-set. A word that is aspirational and confident, not simply wishful. If "utopia" means a world that doesn't exist, I offer the word "verutopia," meaning "a real place," a place we can all work together to build.

It is within our ability to achieve such a world, to bring about the Fifth Age, a world of opportunity and abundance for all. We can build that future, and perhaps even occupy it ourselves. It is no longer simply an idle dream but a real possibility. It is no longer a question of resources, it is simply a question of will.

Acknowledgments

I want to thank my family for their constant support and encouragement throughout the entire process of writing this book. Also, I want to thank my agent and good friend, Scott Hoffman of FolioLit. I suspect he took that first meeting with me as a courtesy to a mutual acquaintance, and that he was as surprised as I was to find in each other a kindred spirit. He is an agent par excellence, and a poker opponent utterly without mercy. Additionally, I must thank my publisher, Peter Borland of Atria. While I am grateful to Peter for publishing this book, I am even more grateful to him for so diligently editing it. He carefully and thoughtfully read through each draft, returning them to me with the kindest of suggestions and only an occasional use of the word "banal." And thanks also to Sean Delone, who waded through the book with Peter. His notes resulted in numerous changes that made the manuscript better.

Next I want to thank Robert Brooker, who offered invaluable suggestions on the manuscript before I sent it off to Peter. Likewise, the lunches I shared with Ellis Oglesby were accompanied by conversations that have influenced many of the pages herein. In addition, I have to give a shout-out to Sunni Brown, who introduced me to Scott. I also want to thank the many people with whom I have had conversations that affected or informed this manuscript: Michael Blend, Shawna Butler,

Nancy Giordano, Jeff Hebert, David Hehman, Gordy Holterman, Brett Hurt, Priscilla Jones, Steve Lanier, Jason Ledlie, Howard Love, Joshua McClure, Kevin Stambaugh, and Stephen Wolfram.

Other people I want to thank who in one way or another have had an impact on this book are Chris Anderson, Joshua Baer, Christina Berry, Christa Haberstock Colson, Jim Connally, Melanie Dunham, Carrin Dunne, Pamela B. Erwin, Peter Handsman, Jason Horton, Steve Baughman Jensen, Fatin Kwasny, Mike Lemper, Drew MacPherson, Patricia Orenthia D. Mason, Ann Knight Meyer, Nancy Mullis, Dave Panos, Ann and Ron Sefrna, Kate Simpson, Meta Jane Smith, Jess Thompson, and Ed Wasser.